Sequential Analysis

There is nothing covered up that will not
be uncovered, nothing hidden that will
not be made known.

Matthew 10:26

Educational, clinical, and social psychologists,
among others, have shown a renewed interest in
what is generally referred to as *sequential
analysis* ... the major impetus behind the resurgence
of sequential analysis methods has been the change in focus
from the study of one organism over time to the study
of the social interaction between organisms. (p. 564)

Dillon, Madden, & Kumar (1983)

Sequential Analysis

A Guide for Behavioral Researchers

JOHN MORDECHAI GOTTMAN
University of Washington

ANUP KUMAR ROY
University of Illinois

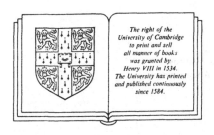

The right of the
University of Cambridge
to print and sell
all manner of books
was granted by
Henry VIII in 1534.
The University has printed
and published continuously
since 1584.

CAMBRIDGE UNIVERSITY PRESS

Cambridge
New York   New Rochelle   Melbourne   Sydney

CAMBRIDGE UNIVERSITY PRESS
Cambridge, New York, Melbourne, Madrid, Cape Town, Singapore, São Paulo

Cambridge University Press
The Edinburgh Building, Cambridge CB2 8RU, UK

Published in the United States of America by Cambridge University Press, New York

www.cambridge.org
Information on this title: www.cambridge.org/9780521346658

A catalogue record for this publication is available from the British Library

Library of Congress Cataloguing in Publication data

Gottman, John Mordechai.
Sequential analysis: a guide for behavioral researchers / John
Mordechai Gottman, Anup Kumar Roy.
p.   cm.
Bibliography: p.
Includes index.
ISBN 0-521-34665-7
1. Social psychology — Methodology. 2. Sequential analysis.
3. Social interaction — Statistical methods. I. Roy, Anup Kumar.
II. Title.
HM251.G695   1989
302'.01'8—dc19   88-38900

ISBN 978-0-521-34665-8 hardback
ISBN 978-0-521-06731-7 paperback

Contents

PREFACE

This book was initially inspired by Gene Sackett's conference on observational methods, held at Lake Wilderness, Washington, in 1976. In many ways that conference was a call for statistical approaches to the detection of interaction sequences in the observation of social behavior. This book is an integration of the methods needed for this task.

Many of the methods were in place prior to 1976, but were not well known. Fortunately, statistical workers have recently shown interest in the problem, and new methods have been developed. In fact, at the time of this writing there may be a bewildering array of procedures for the researcher.

What we present here is an integration of current techniques with suggested guidelines and rules of thumb. We will learn about the power and limitations of these techniques only by using them. Our goal is to make the techniques accessible, and in fact even easy to use.

The book is a sequel to an introductory volume by Roger Bakeman and John Gottman, also published by Cambridge University Press. Beginners should start with that book.

We wish to acknowledge the help of our teachers, particularly James Ringland and Stanley Wasserman. Professor Wasserman was very generous with his time in agreeing to review a draft of this manuscript. This work would not have been possible without the support and release time provided by Research Scientist Development Award K00257MH and a semester's sabbatical leave in 1984 to the first author.

Reading This Book Quickly

We know that some readers will be fascinated by the statistics we discuss, while others will be dismayed. Chapters 1, 2 and 3 of this book are elementary. Chapter 16 is a summary of the major ideas in the book. Section 15.6 is titled "Doing the Minimum," and it contains our most basic set of recommendations that will have reasonably wide applicability.

Part I

Introduction

1

ADVERTISEMENT

This chapter is essentially an advertisement for the sequential and time-series analysis of observational data. We would go so far as to make the statement: *Anyone who has collected data over time and ignores time is missing an opportunity.* We would like to back up this statement with several examples and with a discussion of conceptual issues.

The substantive theoretical discussion in this chapter concerns the analysis of marital and family interaction. Perhaps the most exciting area in which sequential analysis has been applied is social interaction. Sequential analysis, we will see, has supplied a great many analytic tools for researchers studying social interaction. It is not a mere data analysis option. It is a whole new way of thinking about social processes. The dimension of time is so central to conceptualizing social interaction that its use will lead us *to think of interaction itself as temporal form.*

We will begin with a few examples. The first is metacommunication, a concept introduced in what has become known as the "double-bind" paper (Bateson et. al., 1957). A metacommunication qualifies or comments on communication. It can be a nonverbal act that says "all that follows is really play," or it can be a statement that comments on the process of communication, such as, "you're interrupting me," or "that's not what we were discussing." Bateson and his colleagues proposed that the schizophrenic has precisely this deficit, that is, an inability to metacommunicate, which is the way out of the classic double-bind message. In a double-bind message there are two conflicting messages, such as "approach" and "go away." A metacommunication, which would comment on this conflict rather than pursue either alternative, is potentially a solution to the double-bind conflict. Because double-bind messages were hypothesized as characteristic of schizophrenic families, metacommunication was catapulted to a prominent position in the study of social interaction in families. The hypothesis never achieved comparable visibility in the study of marriages.

Gottman (1979a) studied how satisfied and dissatisfied couples differ in the way they try to resolve an important current issue about which they disagree. The discovery was that there was no significant difference in the relative frequency with which satisfied and dissatisfied couples used metacommunication. This was not a sequential analysis. However, the sequential analysis (Gottman, 1979a) showed dramatic differences in the way metacommunication was used by the two groups of couples. Satisfied couples relatively frequently used short chains of metacommunication that functioned as a repair mechanism for the interaction. The metacommunication was usually followed by an agreement by the partner. For example,

A: You're interrupting me.

B: Sorry, what were you saying?

Then the conversation would return to the issue under discussion. The metacommunication was usually delivered with neutral affect, even if the conversation itself had become negative. For dissatisfied couples, metacommunication tended to be delivered with negative affect and followed by a "counter-metacommunication" with the same affect, rather than by an agreement. For example,

A: You're interrupting me.

B: I wouldn't have to, if I could get in a word edgewise.

A: Oh, now I talk too much, is that it?

B: You could say that you do rattle on and on about nothing ...

and so on, almost indefinitely. For dissatisfied couples, metacommunication was like an "absorbing state," to use the language of Markov chains; it was difficult to exit once entered. Furthermore, the affect from the conversation tended to transfer (sequentially) to the metacommunicative chain, so it could not function as a repair mechanism.

Several points should be made about the preceding set of results. First, the sequential analysis revealed differences between the two groups of couples that were not revealed by the analysis that ignored sequence — reason enough to employ sequential analysis. Second, the sequential analysis revealed patterns not even dreamed of by the original paper that drew attention to metacommunication. The implication of this second point is that sequential analysis is an important tool for *generating* theory with good description as well as for testing theory.

This example of metacommunication showed no base rate or unconditional probability differences in metacommunication between couples. Let us now consider an example in which base rate differences do exist and imply a host of differences in social processes, which can only be revealed by sequential analysis. In a monograph on how young children become friends, Gottman (1983) reported that there were large differences between unacquainted children who did and did not "hit it off" (indexed by a questionnaire completed by the mothers about the children's progress toward friendship after the experiment) in the amount of agreement displayed by the guest child. A low level of agreement indexes a staccato rhythm. This rhythm is characteristic of the play of children who do not hit it off. Children who do not hit it off will play for a relatively few number of turns before they escalate by making the play more demanding. They are less likely than children who do hit it off to engage in extended fantasy play, and more likely to have conflict. Here base rate differences between groups do exist and these differences can be *understood* by sequential analyses that illuminate a set of social processes that the base rate differences entail.

A third example is useful for a new reason. Many married couples who attempt to resolve an issue go through a middle phase in which they disagree a great deal. Such disagreement seems to pave the way for later compromise on the issue because they discover the areas of agreement and disagreement in the middle phase. Couples who *avoid* conflict during this middle phase of the discussion often have greater difficulty with compromise, which usually occurs in the last third of the conversation.

Notice how natural this description was, and how rich it is theoretically. It is also consistent with other literature on decision making in areas other than marital interaction. Also notice that this discussion is not even possible without sequential analysis. Furthermore, the sequential analysis needs to be of a certain character, called *nonstationary*. This word is used because the second third of the discussion (in which the couples argue or avoid conflict) is *intrinsically different* from the last third of the discussion, for which the goal is compromise.

Finally, in this advertisement we want to mention the great *conceptual* clarity thinking about social interaction in terms of sequences affords. Just about all of the interesting hypotheses we have about how social systems function imply at their base an imagined scenario of interaction, a scenario invariably sequential in character. In short, since we tend to think sequentially, why not give free rein to this thinking with analytic tools that supplement it?

To illuminate the conceptual clarity to be gained by sequential analysis, let us consider one example from the literature on nonverbal behavior in family interaction, the notion of channel inconsistency, which was also inspired by the double-bind hypothesis. The idea is that one channel communicates a message such as "come hither" while the other channel communicates a message such as "get away from me." The result is supposedly a double bind for the receiver of the inconsistent message. If this hypothesis were true, the *consequences* of the inconsistent message would be predictable. The receiver would act like someone who has been placed in a double bind. On the other hand, it is possible that whenever one channel contains negative affect (e.g., "go away") this is the channel listened to, and the overall message is not at all inconsistent. If anything, the overall message may be interpreted as "this person is negative but trying to qualify, constrain, or temper the negativity." The continuing work of Daphne Bugental in this area supports this latter interpretation. However, a direct test of the hypothesis would be provided by a sequential analysis. This is an example in which theoretical clarity about the anatomy of messages would be gained by thinking in sequential terms.

Another example of the conceptual clarity provided by thinking in sequential terms involves concepts of reciprocity. The reciprocity of self-disclosure has been implicated as the sine qua non of acquaintanceship and close friendship by researchers interested in self-disclosure. In experimental research this reciprocity hypothesis took a temporal form: a self-disclosure by one person would make a subsequent self-disclosure by the other more likely, and, in fact, the intimacy of the self-disclosures would match in more satisfying relationships, or ones in which attraction was high. Despite the temporal nature of this hypothesis it had never been treated sequentially until Dindia's (1981) work that employed Sackett's lag sequential technique (discussed in our Chapter 7). Instead, before Dindia's research, the amounts of self-disclosure by both people were correlated across dyads. This is not a logical test of the reciprocity hypothesis, which by its very nature argues for *contingency*, and hence compares conditional probabilities with unconditional probabilities. The correlation examines only unconditional probabilities. Dindia found no evidence for the reciprocity of self-disclosure using sequential analysis.

Ginsberg and Gottman (1986) studied the importance of the reciprocity of self-disclosure in accounting for variance in the relationship satisfaction of college roommates. Once again, there was no evidence that reciprocity was important.

Why should this be the case? The answer is that so many things happen in a natural conversation other than reciprocal self-disclosure, such as laughing, disagreeing, giving advice, support, and so on, that reciprocal self-disclosure is unusual. In fact, it is difficult to write a script of purely reciprocal self-disclosure that does not seem absurd. No doubt most good friends do find out about one another, but it may, in fact, be important that this mutual discovery be *noncontingent* in many friendships. Self disclosure in children past middle childhood, for example, usually entails support, understanding, and problem solving. Even in adolescents' conversations with their friends in which mutual self-exploration is important, only one problem at a time tends to be discussed (Gottman & Mettetal, 1986).

The importance of the lack of sequential structure in specific interaction situations ought to be noted for family interaction researchers. Lederer and Jackson's (1968) quid pro quo hypothesis of marital functioning is a good example. They proposed that satisfying marriages were characterized by a reciprocal exchange of positive events or actions, a hypothesis that led to a new form of marital therapy called "reciprocal contracting" (e.g., Stuart, 1969).

However, there was never any empirical evidence for the quid pro quo hypothesis, and some evidence (Murstein, Cerretto, & MacDonald, 1977) that the extent to which people held such a philosophy of marriage predicted marital *dissatisfaction*. A series of studies reported in Gottman (1979) found no evidence *within one problem-solving interaction* for positive reciprocity being a variable that discriminated satisfied from dissatisfied couples. Negative reciprocity, on the other hand, did discriminate. Dissatisfied couples were more likely to reciprocate negative affect than were satisfied couples. There was even some evidence for the contention that dissatisfied couples were *more* likely to reciprocate positive affect than were satisfied couples. Gottman proposed the hypothesis that temporal linkage itself was an index of distress.

To understand this result, we can think of the early stages of acquaintanceship. When we first meet someone and are invited for dinner, we want to reciprocate because it shows we are responsive. Similarly, if we are in distressed marriage after a fight, if our spouse is positive we want to reciprocate because it shows we are being responsive. To *not* reciprocate in the context of prior negative affect communicates a great deal. So a positive quid pro quo may make much more sense in unhappy marriages than in happy ones. In a satisfied marriage people tend to be positive fairly independently of their partner's prior actions and based more on their own prior behavior; to use sequential language, autocontingency will probably be more predictive here than will cross-contingency.

Note that the sweeping nature of the original Lederer and Jackson hypothesis has never received an adequate test. Only short time spans have been considered, and primarily problem-solving conversations. It is possible that some form of the hypothesis is true. Our point is that conceptual clarity is added to the hypothesis by thinking sequentially.

We hope we have convinced readers that it would be useful to learn about sequential techniques. The Talmud says, "We are given life. The only question is how to live it." What we are saying is that we are endowed with the ability to think about sequences, the only question is how best to conduct the necessary analyses.

2

HISTORY

This chapter covers the history of sequential analysis, limited primarily to those applications that have bearing on the observation of social behavior in psychology. Readers should not worry too much about undefined terms. We will begin defining terms in Chapter 3.

Initially, interest in the mathematics of sequential analysis had to do with a concern about how to characterize *change*. In a seminal paper in 1952, George Miller introduced Markov processes into psychology by noting that probabilistic methods had proved themselves in sensory psychology and test construction, both of which rely heavily on the independence of observations and on invariance over time. He wrote,

> The basic parameters can be explored at length because sequential effects of measurement are secondary and can be ignored or randomized. (p. 149)

However, for areas of psychology, such as learning, that are essentially concerned with change

> ...it is intrinsic in the very notion of learning that successive measurements are not independent; attempts to use a theory of independent variables must either fail or misrepresent the basic process. Such failures may lead to a rejection of statistical concepts as inadequate; a more proper attitude is to abandon the assumption of independence and ask what help can be had from dependent probabilities. (p. 149)

It is important to note that Miller was writing in the postwar context of the work of Norbert Wiener and his student Claude Shannon, and that the work of both was concerned with change and dependent data.

Before World War II, Wiener had been involved with attempting to model networks of neurons that have some capacity for control and self-regulation. He coined the term "cybernetics" and introduced many engineering terms that eventually made their way into biology and psychology, such as "channel," "feedback," and "homeostasis." Wiener used time-series models to simulate the behavior of neuronal networks.

When the war came Wiener worked on a project of his own invention that also involved these notions of dependency and change. Instead of using flak methods against aircraft, in which it is hoped that a plane will unavoidably hit a piece of flak, Wiener went to work on a gun that could track and anticipate its target. This could only be done by a form of "forecasting" based on some assumption about the past behavior of the target. Thus, information about the *dependent* nature of the data would be used to forecast where to shoot. Time-series analysis was ideal for this task, and Wiener developed many of the central techniques.

Shannon was also involved in some cloak-and-dagger war work in the decoding of secret messages and the cracking of codes. The basic problem was reduced by Shannon to one of communicating a known code (e.g., English) through a noisy channel that lost some information. The key to solving the problem was the discovery of the importance in any code of redundancy, the anatomy of which was central to cracking the code. Terms such as "information," "redundancy," and "structure" were introduced by this thinking. The anatomy of the redundancy in any code is the central subject of sequential analysis.

It is fair to say that initial interest in these techniques in the social and biological sciences was limited to the study of the patterning of behavior within *single* individuals (see Quastler, 1958; Attneave, 1959). Fortunately, however, Gregory Bateson was invited to the 1949 Macy Foundation Conference organized by Norbert Wiener and he immediately took the notions of "communication" to imply communication between people, and notions of "system" to imply interacting social systems such as cultures and families. Unfortunately, although Bateson saw the value of these mathematical tools, he could not apply them.

In a 1976 interview, Bateson and Mead spoke about their reactions to the mathematicians led by Norbert Wiener at the 1946 Macy Conference on cybernetics. Mead said, "All you could ever get out of people like Wiener was 'You need a longer run.' We used to drive them absolutely out of their minds because they were not willing to look at pattern, really. What they wanted was a terribly long run of data." Bateson added: "Of quantitative data, essentially." Neither Bateson nor Mead had any idea how to respond to these requests by the mathematicians. Nor did they see the value in responding.

The classic introductory piece in psychology was Fred Attneave's 1959 monograph *Applications of information theory to psychology.* In this monograph, Attneave introduced the notions of temporal structure. His example had to do with whether someone inventing a random sequence of draws (with replacement) from an urn containing white and black balls is actually inventing a random pattern, or if, on the other hand, there is some temporal patterning to the guesses. For example, if the person has guessed white and then white again, is a black now more likely than it would have been if we knew only that a white had just been guessed? By the way, it turns out that people are not very good at simulating randomness. But this is not the point here. The point is that initial applications of sequential methods had to do with the study of response stereotypy within *one* individual (see Miller & Frick, 1949).

It is interesting that Attneave was apparently unaware of some of the important mathematical papers in the statistics of this field that had emerged since Shannon's classic 1949 work on information theory, particularly Anderson and Goodman's (1957) paper on asymptotic distribution theory of the contingency-like tables derived from Markov matrices.

Actually, the major mathematics for sequential analysis were derived between 1957 and 1962. The Anderson and Goodman paper derived maximum likelihood estimates for transition probabilities in a Markov chain of any order. It also used both likelihood ratio and chi-squared tests for contingency tables to examine the following hypotheses: (1) the transition probabilities of a first-order chain are constant; (2) that when they are constant they are certain specified numbers; and (3) that the process is a chain of a certain order and not another order. The paper developed an asymptotic theory for these inferences when the number of observations increases. The problem of estimating transition probabilities and testing goodness-of-fit and the order of the chain had previously been studied by Bartlett (1951) and by Hoel (1954), in situations where only a single sequence of states is observed, rather than several sequences. The Anderson and Goodman paper generalized these results and provided the basis for the asymptotic statistics used in this book.

Kullback, Kupperman, and Ku (1962) used a statistic called the "minimum discrimination information statistic" (m.d.i.s.) and derived its asymptotic properties. For categorical data that follow a multinomial distribution the m.d.i.s. was the likelihood ratio chi square, which we employ a great deal in this volume. The Kullback et al. paper suggested that m.d.i.s. had the advantage of being capable of "being analyzed into several additive components, similar to the analysis of variance" (p. 576). This work was thus a precursor of log-linear analysis, also employed a great deal in this book.[1]

Two other classic works in the statistics of sequential analysis were Patrick Billingsley's 1961 monograph titled *Statistical inference for Markov processes*, and his 1961 paper in the *Annals of Mathematical Statistics* titled "Statistical methods in Markov chains."

We review the major part of these papers of mathematics later in this book. The idea here is to trace the history of the application of these techniques. The major area first influenced by these new techniques was not psychology, but ethology. Markov models were employed with some success in the 1950s work on the mathematical modeling of learning (e.g., Bush & Mosteller, 1955). Stuart Altmann's 1965 paper in the *Journal of Theoretical Biology* on "stochastics of social communication" among rhesus monkeys must be considered a classic in applying information theory to the study of social interaction. Most of his analysis had to do with the *order* of the Markov model, that is, how far back in the past one could go to gain information in predicting social behavior. Little substantive information was gained by this kind of analysis, however, and Altmann's complex coding system must have been overwhelming to study. He wrote:

> The dyad matrix in this form has been printed. The cells of the matrix list the frequency, relative frequency, and conditional frequency for each of the $(120)^2 = 14,400$ possible pairs, j, k. However, this matrix, which measures 6 x 9 ft, would be difficult to publish; it is available from the author upon request. (p. 503)

---

[1]They also showed that

$$2\sum_i O_i \log (O_i / E_i) \approx \sum_i (O_i - E_i)^2 / E_i$$

by the approximation

$$\log \frac{O_i}{E_i} \approx \frac{1}{2} \frac{O_i^2 - E_i^2}{O_i E_i}$$

where O_i represents observed frequency and $E_i > 0$ represents expected frequency, under some model.

No doubt very few people requested the matrix.

In the same year (1965) Raush published a paper on aggressive children, evaluating Redl and Wineman's life space interviewing program. Using information theory, he concluded that, theoretically, if aggressive and nonaggressive children were mixed in a group, all other things being equal, all children would eventually look like the aggressive children, not like the nonaggressive children. Two years later, using reinforcement theory analysis of children's interaction, Patterson, Littman, and Bricker's (1967) observational study reached a similar conclusion. Raush's work, unlike Altmann's in ethology, was not mainstream in clinical psychology, but it did influence one important researcher, Gerald Patterson, who began studying interaction process in families with aggressive children. Patterson has systematically searched for sequences in families that he calls "coercive processes."

Unfortunately, there were serious problems with the *utility* of many of the early sequential analyses, because they relied on the omnibus tests of information theory (e.g., tests of the order or amount of memory in the Markov process). The questions most researchers have, far more specific than these questions of order, concern whether or not *specific* transition frequencies from an antecedent to a consequent state are different than one would expect if the two states were independent. Subsequent tests (such as chi-squared tests) of specific cells of a transition matrix are possible, but in the past the problem was one of not usually having enough data to employ the tests with confidence.

Within a decade new tests were developed and became widely available in ethology. In fact, there emerged three tools that have begun to be applied in various fields other than ethology: log-linear and logit analysis (see Bishop, Fienberg, & Holland, 1975); time-series analysis (see Gottman, 1981); and lag sequential analysis (see Sackett, 1979). Lag sequential analysis is a technique that must primarily be credited to an extremely creative psychologist, Gene Sackett. The idea is simple, and extremely useful for exploratory data analysis. By computing a sample estimate of sequential connection between two events it is possible to study the strength of the connection at any lag, *ignoring intermediate events.* One can then make inferences about sequences with much less data than would be required for a full Markov chain analysis. In addition, it is possible to study the *pattern* of a relationship between two events as a function of lag to determine such things as whether the relationship is cyclic (see Gottman, 1979a). However, since lag sequential analysis is not a full solution, extreme caution must be used in drawing conclusions about actual sequences.

In the 1960s, many of the influential methodology papers in the field were published in journals of animal behavior (see, for example, Delius, 1969; for summaries of techniques, Chatfield, 1973; Colgan, 1978; Fagen & Young, 1978; Gottman & Notarius, 1978). These papers influenced psychologists who worked with primates (see Sackett, 1978, for example), and these psychologists, in turn, influenced developmental psychologists who studied mother-infant or father-infant interaction (see, for example, Lewis & Rosenblum, 1974).

This activity in the study of social interaction led to three volumes on the analytic options available for observational data (Cairns, 1979; Lamb, Soumi, & Stephenson, 1979; Sackett, 1979), which indicated the great interest that had been aroused in the analysis of observational data using sequential analytic techniques. Several other research areas, among them communications and social psychology, have recently shown an active interest in the sequential analysis of observational data.

As practical experience is gained in employing the range of analytic tools available, an integrated approach to the detection of interaction sequences will emerge. The techniques are now essentially in place, and this book is an attempt to pull them all together in an integrated presentation.

3

THE LANGUAGE OF SEQUENTIAL ANALYSIS

This section will introduce the words commonly used in sequential analysis. Usually, there is a set of coding categories: call them A_1, A_2, $A_3, \ldots, A_S$. For example, in Ekman and Friesen's Facial Action Coding System (FACS), A_1 might be the brows drawn down and together and A_2 might be a nose wrinkle.

The general data situation is illustrated in Table 3.1. Each row represents a code, and each column represents a time interval. The time interval is presumed to be sufficiently small so that coding decisions can be reliably made. Time is used here in a general sense; for most of the techniques discussed in this book all that will matter, to preserve order, is that time 1 precedes time 2, and so on.

One advantage of displaying the data as they are in Table 3.1 is that every code can be represented as a binary time series. Another advantage is that we can add up all codes of a certain type and create a non-binary time series. For example, with the FACS we can add up all codes that indicate brow activity, or all codes that are predictors of sadness. An example of this is given by Tronick, Als, and Brazelton (1977), who graphed the sum of all the behaviors that represented the mother's involvement with her baby and subtracted the sum of all the behaviors that represented the mother's disengagement from the interaction. A similar time series was generated for the baby.

The usual way to display categorical data collected sequentially is to write a string of codes: $A_1 \, A_1 \, A_1 \, A_2 \, A_2 \, A_4 \, A_{12} \, A_1 \, A_1 \, A_8 \, A_8 \ldots$ We would have one such data stream for all subjects. A more complex situation is possible in Table 3.1, in which any number of code categories can co-occur at any time.

Table 3.1. Usual data situation in sequential analysis

			Time				
Codes	1	2	3	. . .	t	. . .	T
A_1							
A_2							
.							
.							
.							
A_i					X_{it}		
.							
.							
.							
A_S							

$X_{it} = 1$ if code i was detected in time interval t
$\quad\;\; = 0$ if code i was not detected in time interval t

3.1 The Moving Time Window

We will illustrate the idea of a moving time window for counting pairs and triples. These are dependent counts: the last member of a pair is the first member of the next pair, and so on. In the sequence AABAB a moving time window two time-units wide would divide the data as follows: (AA)BAB; A(AB)AB; AA(BA)A; and AAB(AB). The parentheses move across the stream of codes and we count the frequency of each pair type (one AA pair, two AB pairs, one BA pair, zero BB pairs). A time window three time-units wide would be used to count triples: (AAB)AB; A(ABA)B; and AA(BAB). Fortunately, this dependency in counting is not a serious problem in statistical analysis thanks to a theorem by Anderson and Goodman (1957). Also see Chapter 9 for a review of a Monte Carlo study on this subject.

3.2 Summary Statistics

There are many ways of summarizing a string of codes such as: AABA-BABBBABAB. There are $n = 12$ observations. We can estimate the probabilities of A and B by their relative frequencies:

$$p(A) = n(A)/n = 6/12 = 0.50$$

$$p(B) = n(B)/n = 6/12 = 0.50.$$

We can also note that A occurs six times, and that of those six times B occurs immediately after A five times. Thus, the conditional probability of B's occurrence, given that A just occurred, is

$$p(B \mid A) \text{ at lag one} = p(B_{+1} \mid A) = 5/6 = 0.83 .$$

Note that the numerator is the frequency of joint occurrence of A and B (number of AB pairs), and that this numerator is then divided by the frequency of A's occurrence. The lag can be greater than one. B is much more likely to occur if we know that A preceded it than it is ordinarily (.83 rather than .50). As noted, to count the number of pairs in a string of codes we use a moving time window of size two time-units:

AABABABBBABAB

The first pair is AA, the second pair is AB, the third pair is BA, and so on. The last element of one pair becomes the first element of the next pair.

Note that we reduce uncertainty in predicting the occurrence of B, or we reduce uncertainty in knowing that the system will next be in state B if we know that A·has just occurred. The probability that B will occur at any time is .50, but if A has just occurred, it is .83. This comparison of conditional with unconditional probabilities and this attempt to reduce uncertainty in prediction by knowledge of prior events are the basis of sequential analysis.

3.3 The Base Rate Problem

Several inappropriate analyses of sequential connection recur frequently in the literature, and it is important to discuss them. The most common is the conditional probability. The idea here is that to establish sequential connection or lag one between antecedent event A and consequent event B, simply examine $\pi(B/A)$, estimated as the frequency of AB divided by the frequency of A. But this statistic does not control for the *base rate problem*, which is that it is meaningless to discuss the conditional probability without reference to the reduction in uncertainty we gain in prediction over and above predicting the base rate of B, $p(B)$. Furthermore, the conditional probability is a poor statistic to use in comparing subjects, because each subject will have a different base rate. For example, a conditional probability of .60 may be unimportant in one subject (because the unconditional is .56) but very important in another subject (because the unconditional is .35).

The second most common inappropriate analysis is a repeated measures analysis of variance with the unconditional probability as one measure and the conditional probability as the other. This error in this analysis is not quite as severe as just using the conditional probability since it is an analysis of the difference between conditional and unconditional probabilities. However, there is no control within-subjects for the variance of this difference based on differences in unconditional probabilities of each code. The variance of the difference is not constant across subjects. We must therefore assume that the variance across subjects of the difference in conditional and unconditional probabilities is adequate. This is like substituting for standard scores the differences between observations and their means without dividing by each cell's estimate of the standard deviation of the mean.

3.4 The Language of Markov Chains

We will now introduce the algebra and language of discrete time Markov chains. This language will be helpful as an approach to thinking about sequence in data.

Assume that a sequence of sample observations of a process is obtained in time. Assume further that the observations consist of a set of mutually exclusive and exhaustive coding categories. Bakeman (1978) described various configurations that produce such data. He called *event sequence data* those data in which events are recorded independent of their duration. For example if we have three codes, O, S, and F, and the observation sample is OOSSSOOFFSFS, the event sequence representation of the sample would be OSOFSFS. The first representation is called *timed event sequence* data. Other data types are possible. For example, Bakeman described *multiple event sequence* data as data in which more than one category at a time is possible (e.g., time 1: mother holds infant, infant vocalizes, mother vocalizes; time 2: mother continues to hold infant, infant vocalizes, mother rocks infant). Each *combination* of codes could be studied in this data representation.

There are two goals in sequential analysis. The first goal of sequential analysis is to discover stochastic (i.e., probabilistic) patterns in the data. The goal is equivalent to "cracking a secret code." We have a large corpus of codes and we wish to determine if these codes have probabilistic patterns of redundancy. In effect, we want to discover the order and the common sequences that characterize the data. The second goal of sequential analysis is to assess the effect of contextual or explanatory variables on sequential structure.

3.5 Markov Models

Let us review. Suppose we observe a system using two codes, A and B. If we observe a sequence AABABABBBABAB, we can describe it nonsequentially by tallying the frequency of A as 6 and the frequency of B as 6. The unconditional probability of A is thus $p(A) = 6/12 = .50$, and the unconditional probability of B is $6/12 = .50$. The conditional probability $p(B/A)$ is computed by noting that B occurs after A five times. Hence the conditional probability of B given A, is $5/6 = p(B/A) = .83$. We thus reduce the uncertainty in predicting state B's occurrence by knowing that the immediately preceding state of the system was A. Thus, the history of the system is useful in predicting the future of the system. This is essentially the line of reasoning we pursue in any sequential analysis.

Some additional examples will help to describe this line of reasoning and provide an introduction to the language of the Markov chain. Jaffe and Feldstein (1970) recorded a 1 whenever a subject was observed to be speaking, and a 0 when silence was observed. They obtained a sequence of ones and zeros: 1111111110001111... Suppose we characterize "a system" as having three possible "states." For example, for dialogues Jaffe and Feldstein defined three states: state 0, both people are silent; state 1, A is speaking; state 2, B is speaking. The fourth possible state of simultaneous speech was so infrequent that it was dropped in some of their analyses.

If a system has three states, 0, 1, and 2, we can describe the system in part by specifying how likely the system is to make transitions between states. The transitions are usually summarized using two matrices, the *frequency state transition matrix* and the *probability state transition matrix*. An example of the frequency state transition matrix is the first-order matrix:

		State at time $t+1$		
		0	1	2
	0	n_{11}	n_{12}	n_{13}
State at time t	1	n_{21}	n_{22}	n_{23}
	2	n_{31}	n_{32}	n_{33}

Each figure n_{ij} is the frequency of transitions from state i to state j. These frequencies can also be converted by dividing each n_{ij} by the row total for row i. These probabilities can be arranged in a matrix called *the probability state transition matrix usually denoted by the letter "P."*

		State at time $t+1$		
		0	1	2
	0	p_{11}	p_{12}	p_{13}
State at time t	1	p_{21}	p_{22}	p_{23}
	2	p_{31}	p_{32}	p_{33}

The rows of this matrix sum to one, and each entry gives the conditional probability that the system, which was in one state at time t, will be in some other specified state at time $t+1$:

p_{ij} = probability of a transition from state i to state j within one time interval

When we restrict our sequential analysis to this state transition matrix, we are claiming that we can characterize the system's movements over time with knowledge of only the immediately preceding event. This claim is that "a first-order Markov model" is adequate for describing the data.

A Markov chain is called "stationary" if this transition probability of going from state i to state j does not depend on when in the chain ($t = 1, 2, 3, ...$) the step is being made. The transition frequencies of a stationary chain are independent of time.

3.6 Example

Gottman and Bakeman (1979) presented an example of a system with four states. The example was taken from the research of Bakeman and Brown (1977) on mother-infant interaction. The four states were: (1) both mother and infant are quiescent; (2) the mother is active; (3) infant is active; (4) mother and infant are coactive. The frequencies of each state were 140, 123, 105, and 68 respectively, out of 436 successive time intervals. Their *frequency state transition matrix* was:

		|	1	2	3	4
		|		*t*+1		
		|	1	2	3	4
	1	|	98	21	14	7
	2	|	21	84	6	12
t	3	|	14	6	60	25
	4	|	7	12	25	24

and the *probability state transition matrix* was:

		|	1	2	3	4
		|		*t*+1		
		|	1	2	3	4
	1	|	.70	.15	.10	.05
	2	|	.17	.68	.05	.10
t	3	|	.13	.06	.57	.24
	4	|	.10	.18	.37	.35

3.7 Higher-order Markov Models

Jaffe and Feldstein in their study of a system with two states yielded the following one-minute sequence of codes:

```
1111111110001111111111111111111111111111111111111111111111
1111111111111111111111000000000000000001111111111111111111
1111111111111111001111111111111111111100111111111111111111111
1111111111110100000001111111
```

This sequence produced the following frequency matrix:

		|	*t*+1		Total
		|	0	1	Total
t	0	|	25	6	31
	1	|	6	162	168
					———
					199

Dividing frequencies by row totals, we obtain the transition matrix:

$$\mathbf{P} = \begin{bmatrix} p_{00} & p_{01} \\ p_{10} & p_{11} \end{bmatrix} = \begin{bmatrix} .81 & .19 \\ .04 & .96 \end{bmatrix}$$

Note again that the rows necessarily sum to 1.0. We can also calculate the *initial unconditional probability vector*:

$$v = \{p(0), p(1)\} = \{.16, .84\}$$

This vector gives the probability that the system will be observed in each of the states.

To summarize (assuming three states), we have an initial state probability vector

$$p^{(0)} = \{p_1^{(0)}, p_2^{(0)}, p_3^{(0)}\}$$

The superscript zeros symbolize the initial $t = 0$ nature of this vector. We also have a one-step conditional probability transition matrix:

$$\mathbf{P} = \begin{bmatrix} p_{11} & p_{12} & p_{13} \\ p_{21} & p_{22} & p_{32} \\ p_{31} & p_{23} & p_{33} \end{bmatrix}$$

Suppose we want to describe the system after n steps, based on the assumption that a first-order Markov model is adequate. Because we know how the system is likely to move from time t to time $t+1$, and we know the initial state vector $p^{(0)}$, we can logically derive the probability vector after n steps, denoted $p^{(n)}$:

$$p_1^{(n)} = p_1^{(n-1)} p_{11} + p_2^{(n-1)} p_{21} + p_3^{(n-1)} p_{31}$$

$$p_2^{(n)} = p_1^{(n-1)} p_{12} + p_2^{(n-1)} p_{22} + p_3^{(n-1)} p_{32}$$

$$p_3^{(n)} = p_1^{(n-1)} p_{13} + p_2^{(n-1)} p_{23} + p_3^{(n-1)} p_{33}$$

These equations, known as the Chapman-Kolmogorov equations, can be written quite neatly in matrix form as $p^{(n)} = p^{(n-1)}$ **P**. They are useful for "rolling forward" the first-order Markov model of the process. With them we can also directly check whether our predicted (first-order Markov) probabilities fit our data.

Jackson and O'Keefe (1982) recently pointed out that there is a misunderstanding of this "rolling forward" idea. Each application of the transition matrix **P** (which can be represented by repeated matrix multiplication) removes us one step from the past. They wrote:

> ...each multiplication of a transition matrix by another transition matrix puts an extra step between the antecedent and the acts being predicted from the antecedent, with the result that there is increased opportunity for something *other than* the antecedent to affect the consequent. (p. 150)

Repeated application of the transition matrix results in a matrix that contains identical rows, each of which is a vector of unconditional probabilities. Because the sequential connections we can detect are usually of low order, as we recede from the past the predictability between the antecedent and the consequent decreases. The fact that the matrix **P** becomes one with equal rows as it is raised to powers simply means:

> Predictions from simple act-to-act connections get worse as
> the distance between predictor and predicted gets larger.
> (Jackson & O'Keefe, 1982, p. 150)

This point has usually been misunderstood, even in our own writing (see, for example, Gottman & Notarius, 1978).

Markov models of higher order than one are possible. In this case, we gain new information by knowledge of events more than one time unit in the past. *Semi-Markov* models (see Hewes, 1979; Howard, 1971) use three sets of parameters, the usual transition matrix, a second matrix whose elements are the probability that a transition will occur between any two states during a particular time interval, and a third matrix whose elements describe the probability that no transition will occur out of the initial state during a particular time interval. The semi-Markov matrix combines events and duration data into one model. The reader is referred to Howard (1971) for a discussion of semi-Markov models.

3.8 The Odds Ratio

In the following table we can compute *the odds* that husbands will be nice as $45/55 = .82$ and the odds that wives will be nice as $60/40 = 1.50$.

	Husbands	Wives
Nice	45	60
Nasty	55	40

The *odds ratio* is the ratio of these two odds, odds $\alpha = .82/1.50 = .55$.

Another statistic commonly computed is the log of the odds ratio. One important property of the odds ratio is its insensitivity to marginal distributions. The second table was obtained by multiplying the first column by ten.

		Marginals				Marginals
55	40	95		550	40	590
45	60	105		450	60	510
		$\alpha = 1.83$				$\alpha = 1.83$

Some writers have argued that the odds ratio is not an ideal statistic because in the following two tables the odds ratio is plus infinity, even though the two kinds of association these tables represent are quite different.

55	0			55	0
0	60			45	60
$\alpha = infinity$				$\alpha = infinity$	

If the general 2×2 table is denoted

n_{11}	n_{12}	n_{1+}
n_{21}	n_{22}	n_{2+}
n_{+1}	n_{+2}	n_{++}

$$\alpha = \frac{n_{11}\, n_{22}}{n_{21}\, n_{12}}$$

$$\beta = \log_e \alpha$$

the variance of α is

$$\sigma_\alpha^2 = \alpha^2 \left(\frac{1}{n_{11}} + \frac{1}{n_{21}} + \frac{1}{n_{12}} + \frac{1}{n_{22}} \right)$$

(all counts n_{ij} are presumed nonzero). The variance of the statistic β (the log of the odds ratio) is:

$$\sigma_\beta^2 = \left(\frac{1}{n_{11}} + \frac{1}{n_{21}} + \frac{1}{n_{12}} + \frac{1}{n_{22}} \right).$$

A statistic analogous to the correlation coefficient is

$$r = \frac{n_{11}\, n_{22} - n_{12}\, n_{21}}{\sqrt{n_{1+}\, n_{2+}\, n_{+1}\, n_{+2}}}.$$

3.9 Four Indices of Sequential Connection

We have stated that the reduction of uncertainty by knowledge of past events is the basic concept in sequential analysis. Can we test the significance of this reduction of uncertainty? Yes, in several ways, four of which are reviewed at this juncture. The remainder of the book is an elaboration on these kinds of techniques, but here we will review a few methods to give the reader an idea of where we are heading.

1. **A binomial test.** This test was first proposed by Sackett (1974), and modified by Gottman (1980) and Allison and Liker (1982). The modified statistic is asymptotically normal; that is, compared to a normal distribution for large $n(n>20)$:

$$z = \frac{p(B_{+k} \mid A) - p(B)}{\sqrt{\dfrac{p(B)\,(1-p(B))\,(1-p(A))}{(n-k)p(A)}}} \tag{3.1}$$

For our example, where $k = 1$,

$$z = \frac{.83-.50}{\sqrt{\dfrac{.50(1-.50)(1-.50)}{(12-1)(0.50)}}} = 2.20$$

which is significant at $\alpha = .05$ because $z > 1.96$. Of course here $n = 12$, not the greater than 20 amount we mentioned, so the significance test may not be very accurate for this example. The k in Equation (3.1) represents cases in which the lag between antecedent and consequent need not be equal to one. Also see Chapter 5 for description of a statistic proposed by Wampold and Margolin (1982).

2.   **Chi-Squared Test.** Using the moving time window we can also compute the *transition frequency matrix*

		Consequent: time $t+1$	
		A	B
	A	1	5
Antecedent: time t	B	4	1

This matrix displays the number of times the system moved from A to A (1), from A to B (5), from B to A (4) and from B to B (1).

As seen earlier, the transition frequency matrix is also sometimes displayed as a *transition probability matrix* by dividing the elements of each row by the row total:

		time $t+1$	
		A	B
	A	.17	.83
time t	B	.80	.20

Note that each row sums to unity. This matrix is sometimes also displayed as a *state transition diagram*. Note also that in this diagram, B would probably be considered almost an *absorbing state*; the system is very likely to move from A to B but not too likely to move back. In a state transition diagram in which the system is very likely to move from A to B but not too likely to move back to A, B would approach being called an *absorbing state*. If it were equally likely to move back and forth, the system might be called "cyclic."

The question is: Are A and B independent? In this case we are comparing $p(B/A)$ to $p(B)$. The answer may be obtained by comparing the observed frequencies to what we could expect in the case of an independence model. As usual, we use the row total times the column total divided by the grand total to estimate the expected frequencies under the independence model.

	Observed		
	A	B	
A	1	5	6
B	4	1	5
	5	6	11

	Expected	
	A	B
A	2.7	3.3
B	2.3	2.7

We compute the Pearson chi square as:

$$X^2 = \sum_{j=1}^{2} \sum_{i=1}^{2} \frac{(O_{ij}-E_{ij})^2}{E_{ij}} = 4.28$$

in which the degrees of freedom are $(s-1)^2$, where s here equals the number of codes; in our case $df = (2-1)^2$.

3. **Likelihood Ratio Chi Square.** We can also compute a statistic that we will have a lot of occasion to use in this book. It is called the "likelihood ratio chi square," denoted LRX^2, or G^2 (after Goodman):

$$LRX^2 = G^2 = 2\sum_{i=1}^{2} \sum_{j=1}^{2} O_{ij} \log \left[\frac{O_{ij}}{E_{ij}}\right].$$

In our case this comes out to $X^2 = 8.30$. The degrees of freedom are still $(s-1)^2$, where $s = $ the number of codes.

4. **Logit Transformation.** We will review a statistical test that also makes it possible to test the significance of the difference in sequential connection across groups. Consider a table in which a wife's antecedent behavior (W_t) takes on one of two values: $W_t = 1$

for negative affect, and $W_t = 0$ for neutral or positive affect. Let us employ a similar notation for the husband's consequent behavior, which we denote H_{t+1}. Consider the following table:

	H_{t+1}	
W_t	1	0
1	109	35
0	35	27

which can also be expressed in terms of proportions as:

	H_{t+1}	
W_t	1	0
1	0.76	0.24
0	0.56	0.44

Bishop et al. (1975) made a fundamental distinction between measures of association in contingency tables that are or are not sensitive to the marginal (row) totals. A measure that is invariant to the marginal totals is provided by the so-called logit transformation (see Allison & Liker, 1982). The logit is defined as:

$$\text{logit}\,(p) = \log_e p/(1-p)\,.$$

We can now define a statistic, called beta, as:

$$\beta = \text{logit}\,[Pr(H_{t+k} = 1 \mid W_t = 1)]$$

$$- \text{logit}\,[Pr(H_{t+k} = 1 \mid W_t = 0)]\,.$$

Note that this statistic is similar to our usual indices of sequential connection that compare conditional probabilities or compare conditional to unconditional probabilities. However, beta has a range from plus to minus infinity. It is zero when the husband's and wife's behaviors are independent.

As we noted earlier, beta is also known as the logarithm of the "odds ratio," which is the cross product ratio in a 2×2 contingency table. If we have a table in which the first row is (a b) and the second row is (c d), then beta is the logarithm of the ratio ad/bc.

A nice feature of employing beta is that we can test to see if beta is different across groups by using a statistic given by Fienberg (1980):

$$z = \frac{\beta_1 - \beta_2}{\sqrt{\sum(1/f_i)}}$$

To illustrate the use of this z-statistic, see Table 3.2, taken from Allison and Liker (1982).

The test in this instance involves comparing the following two betas:

$$\beta_1 = \log_e \frac{76 \times 200}{79 \times 100} = 0.654$$

$$\beta_2 = \log_e \frac{80 \times 39}{43 \times 63} = 0.141$$

The z-statistic is:

$$z = \frac{0.654 - 0.141}{\sqrt{\dfrac{1}{76}+\dfrac{1}{79}+\dfrac{1}{100}+\dfrac{1}{200}+\dfrac{1}{80}+\dfrac{1}{43}+\dfrac{1}{63}+\dfrac{1}{39}}} = 1.493$$

which is not significant at the .05 level (it needs to be larger than 1.96 to be significant).

One analytic option in sequential analysis is to use any of the z-scores for sequential analysis as a statistic of sequential connection without referring the z to a standard normal distribution for significance testing. The z-scores of interest are then computed for each subject, dyad, family, or group in the sample and used in the usual analyses of variance or regressions. An example of this method is a monograph by Gottman (1983) on how children become friends.

Table 3.2. Comparing sequential contingencies across two groups using the log odds ratio.

H_t	Distressed couples W_{t+1}		Nondistressed couples W_{t+1}	
	1	0	1	0
1	76	100	80	63
0	79	200	43	39

3.10 How Children Become Friends

Gottman tape-recorded unacquainted young children and coded these tapes. Specific sequences were identified as indices of the following social processes: (1) communication clarity; (2) information exchange; (3) finding a common ground activity; (4) exploring similarities and differences; (5) conflict resolution; (6) reciprocity of humor, gossip, and fantasy; and (7) self-disclosure. There were three sessions for each dyad. Using six sequential variables in session 2 (z-scores) that indexed information exchange, self-disclosure, the reciprocation of humor, the exploration of similarity with tag questions, and the reciprocation of fantasy in a stepwise multiple regression analysis, Gottman was able to account for 92.3% of the variance in mothers' questionnaires of the children's progress toward friendship two months after the study was completed.

3.11 Discrete Time Semi-Markov Models

Semi-Markov models integrate timed data and event data in one analysis. This section will briefly introduce the sort of parameters in the discrete time semi-Markov process, with an example of such an analysis. We should use semi-Markov models only if we believe that transitions to another particular state j are a function of the time spent in an antecedent state i. Thus, the usual transition probability matrix in Markov models ought to be qualified in some way by the time the system has been "waiting" in a state. If a particular transition becomes *less* likely with a longer wait, the system is said to have "inertia" (see Bartholomew, 1982).

We compute *interval transition matrices*, which are a function of both the particular transition of interest, $i\ j$, and how long ago, k time units, the transition occurred. The values in the matrix are $p_{ij}(k)$, the probability that the system will be in state j given it entered j from i, k time units ago. For more information, see Howard (1971).

Bowe and Anders (1979) used a semi-Markov model to describe the sleep-wake states of infants 2 and 9 months of age. They used time-lapse videorecorders sensitive to infrared light and coded one of four states: (1) quiet sleep (QS); (2) active sleep (AS); (3) awake (AW); (4) out of crib (OOC). They compared the two age groups for each third of the night, using analysis of variance to compare transition probabilities and holding times.

They compared two sets of parameters: O_{ij}, the probability of being in state j at time t, given a start-in state i; and the *holding time*, defined as the amount of time the infant stays in state i, given that the next transition will be to state j. Holding times for the 2-month-old infants were almost twice as long as for 9-month-old infants in AS before a QS transition in the early night. Also,

> In the middle of the night, both age groups have similar holding times in AS before a QS transition, but the older infants have significantly longer QS holding times before an AS transition. This tendency for longer QS holding times for the older group is present throughout the night. Finally, the younger group is more likely to transition to AW. That is, their AS → QS sleep cycling is more easily disrupted. (p. 45)

3.12 Our Overall Plan

In this book we will discuss sequential analysis in two steps. The first involves fitting some Markov model to a set of data concerning one subject, couple, family group, and so on. We will refer to this as *fitting the timetable*. Issues of fitting a timetable involve determining the order of the Markov chain and stationarity.

The second step is to determine how the timetable changes as a function of the experimental design. For example, we may be studying marital interaction and have the following events: Husband positive affect (H+), Husband negative affect (H−), Husband neutral affect (H0),

Wife positive affect (W+), Wife negative affect (W−), and Wife neutral affect (W0). The timetable could be a stationary second-order Markov process. This implies that it is meaningful to go two events back in time to reduce uncertainty in our ability to predict the probability of an event.

If we were studying marital interaction in blue-collar and white-collar couples who were either happily or unhappily married, we would have a 2×2 factorial design as a contextual design. Each couple in each cell would have a second-order Markov transition frequency matrix, or "timetable." We would probably like to know how the timetable varies as a function of social class, and of marital satisfaction, and the interaction of these two contextual variables.

Issues involved in analyzing the contextual design include whether to analyze the data subject by subject using analysis of variance on regression techniques, or to pool across subjects and use log-linear. An issue of pooling is whether there is homogeneity in the timetables across subjects. Logit models can also be used without pooling. We will discuss all these issues in this book. To summarize, we have discussed four options so far.

Option 1. Do not pool data. Compute indices of sequential connection for each subject (dyad, etc.) in the study. Do standard ANOVAS, regressions, etc. References: Gottman (1983); Bakeman and Brown (1980).

Option 2. Pool data across subjects. Test timetable's order, stationarity, homogeneity. Then see if the timetable varies with the contextual variables in our experimental design. References: Krokoff, Gottman, and Roy (in press), and Chapter 11, this volume.

Option 3. Study the timetable for each subject (dyad, etc.) in the study. Useful for exploratory analyses. Reference: Chapter 12, this volume.

Option 4. Create time series from categorical data. Reference: Bakeman and Gottman (1986).

Part II

Fitting the Timetable

4

THE ORDER OF THE MARKOV CHAIN

Originally, information theory was devised in response to a need to conceptualize the amount of information in a message that might be transmitted through a noisy channel. In this event errors would clearly take place in transmitting the message, and they would be more or less serious depending on the amount of *redundancy* in the language system used. In English, for example, if we received the word "Q_ICK" we could guess with complete confidence that the missing letter was "U." We can make these guesses because languages have a *temporal structure*. What does this temporal structure mean? It means simply that we gain information in prediction by knowing the past.

> *Temporal structure means that we gain information in prediction by knowing the past.*

There was also a great deal of concern during World War II about breaking secret codes that the Axis powers used to transmit information. Conceptually the code-breaking problem is very similar to the problems we have in analyzing a sequence of codes. We inductively seek to determine the sequential structures that are present in the data.

4.1 Shannon's Approximations to English

In his 1949 paper, Claude Shannon generated a series of approximations to English. He used a 27-symbol alphabet: 26 letters and a space. The first approximation assumed that all symbols were independent and equally probable; these obviously not very good assumptions produced

XFMOL_RXKHRJFFJUJ_ZLPWCFWKCYJ_FFJEYV
KCQSGHYD_QPAAMKBZAACIBZLHJQD

which looks very little like a typical sample of English text. Shannon's next approximation was to *account for asymmetries in the frequencies of the symbols.* For example, obviously "A" occurs more frequently in English than "X." The next approximation produced

OCRO_HLI_RGWR_NMIELWIS_EU_IL_NBNESEBYA
TH_EEI_ALHENHTTPA_OOBTTVA_NAH_BRL

This process was next continued by accounting for rules between two immediately adjacent symbols, called *digram structure* by Shannon. For example, the "U-follows-Q" rule is an example of digram structure. However – and this is an important point – the digram structure is usually probabilistic. That is, for most English letter pairs, usually if there is a structure this means there is an increased or decreased *probability* that the second letter will follow the first, compared to its usual probability of occurrence. We are not speaking of certainties when we speak of digram structure.

> *Digram structure between two symbols means that the second symbol has a changed probability of occurrence when it follows the first, compared to its usual probability of occurrence.*

This point is critical because it implies that we must always compare the conditional probability of occurrence (the probability of the second symbol in all those pairs in which the first symbol occurred) to the unconditional probability of occurrence of the second symbol.

We will often return to these two points: the probabilistic nature of temporal structure, and the comparison of conditional and unconditional probabilities.

Shannon's next approximation to English took its digram structure into account. It was, in part,

ON_IE_ANTSOUTINYS_ARE_T_INCTORE_
BE_SDEAMY_ACHIN_D_ILONASIVE.

This piece of text begins to resemble English a bit. Shannon next included the *trigram structure* of English. It is critical to keep in mind that when we refer to trigram structure, we mean *new* information is obtained by going back into the past an additional unit for the second preceding symbol. The key word here is *new* information. The third order (trigram) approximation was

IN_NO_IST_LAT_WHEY_CRATICT_FROURE_
BIRSGROCID_PONDENOME_OF_DEMONSTURES_
OF_THE_REPTAGIN_IS_REGOACTINA_OF_
CRE.

It is fascinating that Shannon's next example moved from the letter to the word as a unit and began with digram, trigram, and so on, structure among *words*. It will forever be a mystery why Shannon abandoned the letter unit in this example; he said only that "it is easier and better to jump at this point to word units" (p. 43). The question is, could Shannon have created approximations to English that were as good if he had stuck to the letter as the unit? This question has profound implications for the work of sequential analysis.

4.2 Information

Shannon began discussing sequential dependency by talking about redundancy, and he did this by introducing the concept of "information." We will start with an example of a set of sequential data and work through all the basic concepts with this one example. Of necessity this will be a simple example, actually rather trivial because we will assume a sequence generated by only two codes. We refer to the two codes as A and B. For example, code A may represent the Facial Action Code (Ekman & Friesen, 1978) AU9, the Nose Wrinkler, and AU10, the upper-lip raiser. Both facial action units may be involved in a general-disgust facial expression of emotion, and we may wish to examine the issue of whether they are indeed temporally related. So we examine a filmed record for these two codes and obtain the following data out of a total of 44 observations:

ABABABAAABBABABABAABBABABABAAB
BABABABABBBAAA

This is our raw-data record.

4.3 The Plan of Our Discussion

The plan in this section is first to explain the computations, that is, how to compute the necessary statistics and how to come to conclusions about the data. We do this in two ways, one involving information theory, and the other involving the likelihood ratio chi square. The information theory computations, important only for historical and heuristic reasons, well serve to illustrate the types of computations we will eventually discuss. Readers do not need to learn how to perform these information theory statistics, but the discussion will add something conceptually to their understanding of sequential connection.

The data for now are the record for one unit of interacting subjects. Later we will take up issues of experimental design. After this pass through the practical computational world, we will investigate the intuitive side of the statistics, that is, why *these* particular statistics may work. Readers should keep this plan in mind to avoid frustration.

Here is a very fundamental principle in our analysis of these data:

All analyses will be concerned with searches for asymmetries.

For example, Shannon, in his first empirically based approximation to English, took into account asymmetries in the frequency of occurrence of the English letters. In our example, we count how often A and B occur and we obtain 23 A's and 21 B's. They both seem to occur about as often.

4.4 Likelihood Ratio Chi-Squared Tests

In this chapter we will not use the familiar Pearson chi-squared test because it will be very useful to introduce a second kind of chi square that is called the likelihood ratio chi square (LRX^2). The equations for the Pearson X^2 and LRX^2 are somewhat similar. Both require expected counts under some model, the E_i, and observed counts, the O_i. The equations are:

$$\text{Pearson } X^2 = \sum (O_i - E_i)^2 / E_i \tag{4.1}$$

$$LRX^2 = 2\sum O_i \log_e (O_i/E_i) \tag{4.2}$$

Both chi-square statistics will be useful to us. See Appendix 4.2 of this chapter for a brief discussion of Equation (4.2).

Equation (4.2) is very important because almost all our tests in this book will be based on some comparison of observed (O_i) to expected (E_i) frequencies. There are many ways to derive the E_i, and we usually do so from some set of logical assumptions called *a model*. In fact, for our purposes a model is a way of getting a set of E_i for comparison with the set of O_i.

Equation (4.2) is sometimes written in terms of observed frequencies, $(n_{observed})_i$, and observed and expected probabilities $(p_{observed})_i$, $(p_{expected})_i$ as

$$LRX^2 = 2\sum_i (n_{observed})_i \log_e \frac{(p_{observed})_i}{(p_{expected})_i} \qquad (4.3)$$

How would we statistically *test* whether there was an asymmetry in the frequency of occurrence of A and B of a fixed total of 44 observations? We could use a likelihood ratio chi-square test. If A and B were expected to occur equally often, the E_A and E_B, the expected frequencies would be $44/2 = 22$. Our observed frequencies, O_A and O_B, were 23 and 21 respectively, and we know that

$$LRX^2 = 2\sum O_i \log_e (O_i/E_i)$$

$$= 2(23 \log_e (23/22) + 21 \log_e (21/22))$$

$$= 2(1.02 - 0.98) = 0.09 ,$$

which is not statistically significant, so that we cannot reject the model that led us to expect A and B to occur equally often.

Remember that:

DEFINITION OF A MODEL:

A "model" is any set of assumptions or any reasoning at all, for that matter, that would lead us to generate some expected values.

We really can get the E's any way that is reasonable and specifiable. However, we must be cautious because LRX^2 will only be asymptotically chi-square if the E's are maximum likelihood estimates.

4.5 Digram Structure

What we must do next is to employ the moving time window to count how many pairs, out of 43 pairs, we have of each type. In so doing we will discover that we have 6 AA pairs, 16 AB pairs, 16 BA pairs, and 5 BB pairs. Recall that there were 23 instances of A and 21 instances of B. So if there were no digram structure whatsoever, what would we have expected? The probability of A occurring at any point in time is $23/44 = 0.52$, and of B occurring at any point in time is $21/44 = .48$. Because there are 23 instances of A, we would expect $23 \times .52 = 12$ AA pairs, $23 \times .48 = 11$ AB pairs, $21 \times .52 = 11$ BA pairs, and $21 \times .48 = 10$ BB pairs. What would be the value of our chi-square statistic?

$$LRX^2 = 2\sum O_i \log_e (O_i/E_i)$$

$$= 2(6 \log_e (6/12) + 16 \log_e (16/11) + 5 \log_e (5/10) + 16 \log_e (16/11))$$

$$= 2(-4.16 + 6.00 - 3.47 + 6.00) = 8.75$$

which is significant at one degree of freedom ($p < .05$). This shows us that we do have a digram structure of some sort.

4.6 Contingency-Style Tables

As we have seen, categorical data collected over time can be arranged in a manner similar to a contingency table. In this section new notation will be introduced. Suppose that there are two codes (or categories, or states) denoted 1 and 2. If the data stream is 112112, then the first pair is 11, the second is 12, the third is 21, the fourth is 11, and the fifth pair is 12. Note once again that pairs are not independent because the second state of one pair is the first state of the next pair. Although we use the term "contingency table" in this book, this table is actually not the familiar contingency table we might obtain if we sampled 1,000 people to determine if they were male or female, and Republican or Democratic. However, the data still consist of counts or frequencies, and we can array them in a two-dimensional frequency table (see Table 4.1).

Table 4.1. General notation and specific example of frequencies in a 2-dimensional table

		Consequent	Code	Marginals
		1	2	
Antecedent Code	1	n_{11}	n_{12}	n_{1+}
	2	n_{21}	n_{22}	n_{2+}
Marginals		n_{+1}	n_{+2}	N_2

N_2 = the number of pairs

In this notation

n_{11} = the number of (1, 1) pairs

n_{12} = the number of (1, 2) pairs

n_{21} = the number of (2, 1) pairs

n_{22} = the number of (2, 2) pairs .

If the data stream is 11112212122211, the pairs are 11, 11, 11, 12, 22, 21, 12, 21, 12, 22, 21, 11. Thus, $n_{11} = 4$, $n_{12} = 3$, $n_{21} = 3$, $n_{22} = 2$, $n_{1+} = 7$, $n_{2+} = 5$, $n_{+1} = 7$, $n_{+2} = 5$. The plus represents the subscript we sum across. Note that N_2, the number of pairs, is 13, while the number of observations, denoted N_1, is 14. In general, in one data stream, $N_2 = N_1 - 1$. Also, the number of triples $N_3 = N_2 - 1$, and so on. This fact is called an "edge effect." If the data consists of k streams then $N_1 = N_2 + k$.

We will now employ the clear notation Chatfield (1973) suggested for probabilities. We denote: $p(i)$ = the probability of state i, which can be estimated from the data in a variety of ways: $p(i,j)$ = the joint probability of two "successive" states, the first being state i and the second state j. We can estimate $p(i,j)$ by

$$p(i,j) = n_{ij} / N_2 .$$

The denominator is the total number of pairs observed. In a similar way, we denote: $p(i,j,k) =$ the joint probability of the triple (i,j,k). The denominator in the estimate $p(i,j,k) = n_{ijk} / N_3$ is the number of triples observed. We denote $p(j/i) =$ the conditional probability of state j, given that the previous state was i. This is usually estimated as the proportion of times that when i occurred, j followed immediately. The estimate usually is

$$p(j/i) = n_{ij} / n_{i+} .$$

Referring to Table 4.1, this fraction is the number of (i,j) pairs divided by an estimate of n_i; note that $n_{i+} \neq n_i$ because of edge effects.

In a similar manner we denote $p(k/i,j) =$ conditional probability of state k given that the two previous states were i and j, where i immediately preceded j. The usual denominator of $p(k/i,j)$ is the number of (i,j) pairs, estimated as n_{ij}.

4.7 Independence

Antecedent and consequent will be independent if

$$p(i,j) = p(i) p(j) \text{ for all } i \text{ and } j .$$

or equivalently, if:

$$p(j/i) = p(j) \text{ for all } i \text{ and } j .$$

This latter equation states that antecedent and consequent are independent if the conditional and unconditional probabilities are equal. As we have noted, this is also called a "zero-order Markov model," assuming that no dependencies exist among other time lags. At most a first-order Markov model holds if:

$p(k/i,j) = p(k/j)$ for all i, j, and k .

4.8 Testing Independence

Recall that Table 4.1 is not actually a contingency table, in that successive pairs of observations are not independent because the second state of one pair is the first state of the next pair. However, it has been shown (Anderson & Goodman, 1957) that as N_1 becomes large, the LRX^2 statistic is distributed like a chi-square random variable, under the null hypothesis that the independence assumption is correct. For the antecedent-consequent case (Table 4.1), if successive observations are independent, the expected number of times that state i will be followed by state j is

$$e_{ij} = N_2\, p(i)\, p(j)$$

The probabilities $p(i)$ are estimated from the data by n_{i+} / N_2 and n_{+j} / N_2, respectively, and thus,

$$\hat{e}_{ij} = n_{i+}\, n_{+j}\, / N_2 .$$

In this case, under the null hypothesis of independence the following variable is distributed as chi square

$$LRX^2 = 2\sum_{i,j} n_{ij}\, \log\,(n_{ij}/\hat{e}_{ij})$$

with $(c - 1)^2$ degrees of freedom, where c = the number of categories (or states) in the coding system. This is a statistical test of whether the system is a first-order Markov process.

A similar statistic applies for testing whether the process is a second-order Markov process. The observed frequencies of triplets are compared to the expected frequencies, under the assumption that the process is actually a first-order Markov process. The expected frequencies are

$$e_{ijk} = N_3 \, p(i,j,k)$$

$$= N_3 \, p(i,j) \, p(k/i,j) \; .$$

This much is an identity. Now if we assume that the process is a first-order Markov chain, then

$$p(k/i,j) = p(k/j) \; ,$$

so that

$$e_{ijk} = N_3 \, p(i,j) \, p(k/j) \; .$$

These are the frequencies we would expect under a first-order Markov assumption. These two probabilities can be estimated from the data by $p(i,j) = n_{ij+} / N_3$, and $p(k/j) = n_{jk+} / n_{j++}$. The latter estimate is an estimate of the number of n_{jk} pairs, divided by an estimate of the number of times j was observed. Thus,

$$\hat{e}_{ijk} = n_{ij+} \, n_{jk+} / n_{j++}$$

the test statistic is

$$LRX^2 = 2 \sum_{i,j,k} n_{ijk} \log \left(n_{ijk} / \hat{e}_{ijk} \right)$$

with $c(c-1)^2$ degrees of freedom. Once again, the test statistic is only asymptotically distributed as chi square.

4.9 Cellwise Examination of Contingency Tables

Once we know that there is a significant chi-squared test statistic (i.e., the independence model has been rejected), we need to know what cells create the observed effect. This is similar to subsequent tests in the analysis of variance. Of the five options available, one is to partition the contingency table, examining components of the Pearson X^2 statistic,

Table 4.2. Example of the third procedure for cellwise comparison after the independence model is rejected

	Observed				Expected	
	B_1	B_2			B_1	B_2
A_1	1	5	6	A_1	3.9	2.1
A_2	12	2	14	A_2	9.1	4.9
	13	7	20			

$\sum (O - E)^2 / E$. We will not discuss this option in this book, but see two papers: Castellan (1965), and Bresnahan and Shapiro (1966).

The second procedure is to examine the cell probabilities under the assumption of the rejected null independence model. For the binomial distribution, we can use Equation (4.1), Chapter 4.1.

We must be cautious in cellwise examination because of the multiple comparison problem. The solution is to adjust alpha accordingly. We can also use $\sqrt{X^2_{df}/df}$ as the comparison statistic instead of $N(0,1)$, but this is not a very sensitive alternative. Fagen and Mankovich (1980) reviewed several other such statistics, and argued that they are equivalent.

The third procedure is to employ Freeman-Tukey deviates. We compare the statistic

$$F.T. = \sqrt{\text{observed}} + \sqrt{\text{observed} + 1} - \sqrt{4(\text{expected}) + 1} \qquad (4.4)$$

with an $N(0,1)$ distribution. This result is derived from the variance stabilizing transformation for Poisson data. In the $A_2 B_1$ cell of Table 4.2

$$\sqrt{12} + \sqrt{13} - \sqrt{37} = .99 \ ,$$

which is not significant.

The fourth procedure is to examine components of the likelihood ratio statistic, and to compare this with a chi-square distribution with degrees of freedom $= df /\#$ (cells). We defer the major discussion of the

Table 4.3. Frequencies of n-grams, showing specific patterns of asymmetry

Quartogram	n_i(quart)	Trigram	n_i(tri)	Digram	n_i(di)	Symbol	n_i(sym)
AAAA	0	AAA	2	AA	6	A	23
AAAB	1						
AABA	0	AAB	3				
AABB	3						
ABAA	3	ABA	12	AB	16		
ABAB	9						
ABBA	3	ABB	4				
ABBB	1						
BAAA	2	BAA	4	BA	16	B	21
BAAB	2						
BABA	11	BAB	12				
BABB	1						
BBAA	1	BBA	4	BB	5		
BBAB	3						
BBBA	1	BBB	1				
BBBB	0						

LRX^2 statistic until Chapter 10. Fifth, Fagen and Mankovich (1980) reviewed Bishop et al. (1975, pp. 140-141), who suggested computing LRX^2 for the whole table, then repeating but omitting the cell in question (i.e., treat the cell as a structural zero). The difference in the two LRX^2's is chi square with one degree of freedom. Fagen and Mankovich (1980), however, reviewed and recommended Brown's (1974) method for detecting "multiple" outlier cells. Brown's stepwise method ranks these cells in order of significance; the procedure is, however, specific to two-way contingency tables. This fifth recommendation will become clearer in Chapter 10, where we will also present a specific procedure for examining contrasts.

4.10 Information Theory

We can use moving time windows of various sizes to construct Table 4.3. The test we are really interested in is not whether there is trigram or quartogram structure, but whether there is "additional" structure every time we go back one more symbol into the past. Obviously if there is asymmetry in the pair structure (as we discovered there is), there will be asymmetry in the trigram frequencies. It stands to reason that if there are more AB than AA pairs, there will be more ABA triples than AAA triples. But that is boring. We want to know if we will encounter new surprises by going back an additional time unit into the past. Hence, when we ask about trigram structure we want to control for digram structure.

To accomplish this we have to compute a set of numbers called T_i, *which tell us how much information is gained by basing predictions on the previous i events rather than on the previous (i − 1) events.* These numbers T_i are computed as

$$T_i = H_i - H_{i+1}$$

where the numbers H_i are computed as follows:

1. We will first compute H(symbol), H(digram), H(trigram), H(quartogram), etc.

2. Next we compute the H_i, defined as
   $H_i = H(N-gram) - H(N-1gram)$

3. Then we compute the T_i, defined as above.

   a. **Computing H(symbol).** If there are c symbols in our coding system, and if the ith symbol occurs with frequency n_i, and there are a total of N_1 observations, then H(symbol) is computed as follows:

   $$H(symbol) = (\log_2 N_1) - \frac{1}{N_1} (\sum_{i=1}^{c} n_i \log_2 n_i) \qquad (4.5)$$

   For Table 4.3, $N_1 = 44$, $n_1 = 23$, $n_2 = 21$, and $\hat{H}$(symbol) is $5.46 - (1/44)(196.34) = .998$. This number is the average amount of information contained in a symbol.

   b. **Computing H(digram).** If there are N_2 total pairs and if pair i has frequency n_i, then the average information in pairs of events is

   $$H(digram) = (\log_2 N_2) - \frac{1}{N_2} (\sum_{i=1}^{c^2} n_i \log_2 n_i)$$

Table 4.4. Worked example

Quartogram	n	$n \log_2 n$	Trigram	n	$n \log_2 n$	Digram	n	$n \log_2 n$	Symbol	n	$n \log_2 n$
AAAA	0		AAA	2	2.01	AA	6	15.51	A	23	104.07
AAAB	1	0.00									
AAAB	0		AAB	3	4.76						
AABB	3	4.76									
BBAA	3	4.76	ABA	12	43.03	AB	16	64.01			
ABAB	9	28.54									
ABBA	3	4.76	ABB	4	8.01						
ABBB	1	0.00									
BAAA	2	2.01	BAA	4	8.01	BA	16	64.01	B	21	92.27
BAAB	2	2.01									
BABA	11	38.07	BAB	12	43.03						
BABB	1	0.00									
BBAA	1	0.00	BBA	4	8.01	BB	5	11.62			
BBAA	3	4.76									
BBBA	1	0.00	BBB	1	0.00						
BBBB	0										
Σ	41	89.67		42	116.85		43	155.15		44	196.34

$\hat{H}_0 = 1$

$\hat{H}(symbol) = \log_2 44 - (\frac{1}{44})(196.34) = .998$

$\hat{H}(digram) = \log_2 43 - (\frac{1}{43})(155.15) = 1.822$

$\hat{H}(trigram) = \log_2 42 - (\frac{1}{42})(116.85) = 2.611$

$\hat{H}(quartogram) = \log_2 41 - (\frac{1}{41})(89.67) = 3.172$

$\hat{H}_1 = \hat{H}(symbol) - \hat{H}_0 = -.002$

$\hat{H}_2 = \hat{H}(digram) - \hat{H}(symbol) = .824$

$\hat{H}_3 = \hat{H}(trigram) - \hat{H}(digram) = .789$

$\hat{H}_4 = \hat{H}(quartogram) - \hat{H}(trigram) = .561$

For Table 4.3, $\hat{H}$(digram) is $5.43 - 1/43(155.15) = 1.82$. We can continue in this fashion, as illustrated in Table 4.4. The important computations are the T's because Attneave (1959) noted that

$$X^2 = 1.3863 K T_n \quad (df = c^{n-1}(c-1)^2) \qquad (4.6)$$

where K is the total number of observations (45). First we need to compute the $\hat{H}_i$, as shown in the bottom of Table 4.4. Once we have computed the $\hat{H}_i$, we can compute the $\hat{T}_i$, as follows:

$$\hat{T}_1 = \hat{H}_1 - \hat{H}_2 = 0.822 \text{ and } X^2 = 5.752, \quad df = 1$$

$$\hat{T}_2 = \hat{H}_2 - \hat{H}_3 = 0.035 \text{ and } X^2 = 0.245, \quad df = 2$$

$$\hat{T}_3 = \hat{H}_3 - \hat{H}_4 = 0.228 \text{ and } X^2 = 1.595, \quad df = 4 .$$

These computations show us that only basing predictions on the immediately previous event gives us any significant information, in other words, only T_1 is significant. The data have only a digram structure.

c. **Summary**. Using moving time windows of varying length, a table like Table 4.3 is completed. The important statistics are the T_i, which are distributed as shown in Equation (4.6), and which assess whether each step into the past provides new information.

We will now discuss a bit of the rationale for these computations and discuss the rationale for the definition of H.

4.11 Reduction in Uncertainty

Suppose we want to guess the location of a specific target square out of the 64 squares on a checkerboard, and we can ask only yes-no questions. How many questions would be necessary to locate the target square? The answer is, six questions. The first would ask something like "Is it one of the 32 squares on the left hand of the board?" The answer reduces the number of alternatives in half. The next question could be "Is it one of the 16 in the upper half of the remaining 32 squares?" The answer again reduces the number of remaining alternatives in half.

Each answer furnishes *one* bit, a binary unit of information. Thus, the uncertainty to the question "What square of the checkerboard am I thinking of?" amounts to six bits of information.

How can one quantify this unit of information? We note that it is equivalent to saying that if each alternative is equally likely, the "information" is equal to the logarithm to the base 2 of the number of alternatives. Sixty-four squares is 2^6, and the number of bits is $\log_2(2^6) = 6$. This was the number of yes-no questions we had to ask to locate our target square. The information can thus be defined as:

$$H = \log_2 m$$

where m is the number of equally probable alternatives. If each alternative has probability $p = 1/m$, this means

$$H = \log_2 (1/p) = -\log_2 p$$

When the alternatives are *not* equally likely, the formula can be generalized to a weighted average as follows. If each of i alternatives has probability p_i, then

$$H = \sum_i p_i \log_2 (1/p_i)$$

For example, if we had a biased coin that came up heads 90% of the time, then the information associated with a throw of this coin is

$$H = \sum_i p_i \log_2 (1/p_i)$$

$$= 0.90 \log_2 (1/0.90) + 0.10 \log_2 (1/0.10)$$

$$= (0.90)(0.15) + (0.10)(3.32) = 0.47 \text{ bits}$$

Note that information associated with a throw of an unbiased coin is

$$H = 0.50 \log_2 (1/0.50) + \log_2 (1/0.50)$$

$$= (0.50)(1.00) + (0.50)(1.00) = 1.00 \text{ bits}$$

There is more information in the unbiased case, because the toss of the unbiased coin is more uncertain than the toss of the biased coin. This may be counterintuitive, in its implication that certainty means no information. Uncertainty is an important concept in information theory.

Suppose a sequence of codes is redundant to some order: that is, it contains unknown sequential dependencies. The first-order estimate of H, the amount of information in the sequence is:

$$\hat{H}_1 = \sum_i \hat{p}_i \log(1/\hat{p}_i)$$

This number assesses asymmetry in the frequency of the different codes.

The second-order estimate H_2 considers the possibility that there is some information in *pairs* of consecutive codes, which is calculated as if each *pair* was a separate code.

$$\hat{H}(\text{pairs}) = \sum_i \hat{p}(\text{pair}_i) \log_2(1/\hat{p}(\text{pair}_i))$$

The second-order estimate of H takes the difference between $\hat{H}(\text{pairs})$ and $\hat{H}_1$:

$$\hat{H}_2 = \hat{H}(\text{pairs}) - \hat{H}_1$$

This difference assesses the asymmetry in pairs of codes, independent of the possibility that they are not all equally probable. If the quantity is nonzero, this is equivalent to the requirement that the conditional probabilities exceed unconditional probabilities, a common test in sequential analysis. Such a requirement means that knowledge of one event A reduces uncertainty in the occurrence of another event B, compared to a prediction that would be made simply by knowledge of the unconditional frequency of B, that is,

$$p(B \mid A) \neq p(B)$$

Continuing in this vein, H_3 is the difference between H(triplets) and H(pairs). The sequence H_0 (which is $\log_2 C$, where C is the number of coding categories, which in our case equals 2), $H_1, H_2, H_3, \ldots$, is a decreasing sequence that measures the conditional uncertainty for each order of dependence. Chatfield and Lemon (1970) wrote that it is normally unnecessary to compute more than six of these quantities before they become unimportant. The quantities $(2 \log_e 2 N_{i+1})\hat{T}_i$ become asymptotically chi square (recall that N_i is the number of i-tuples, using the moving time window) with degrees of freedom $c^{i-1}(c-1)^2$ for $i > 0$, and $c - 1$ for $i = 0$. Chatfield and Lemon (1970) wrote that this result follows from the appropriate likelihood ratio test (Hoel, 1954). Chatfield (1973) wrote that the statistic is the same as the likelihood ratio statistic for testing the null hypothesis that the sequence has sequential structure of order $(i - 1)$, not order i. The asymptotic chi-squared test Chatfield referred to was derived by Bartlett (1951) for the situation in which the transition probabilities are known and not estimated. Hoel (1954) extended Bartlett's results to the case in which the transition probabilities are not known but have to be estimated from the data. Attneave (1959) uses the quantity $2 \log_e 2 \; K\hat{T_i}$ where K is the total number of observations. These two numbers are not very different and will, of course, be equivalent asymptotically. Attneave's is easier computationally.

4.12 Exercise

Readers who wish to understand these computations more thoroughly can gain in confidence if they work through this example.

Attneave (1959) gave the example of a series of 203 codes as shown below.

```
WBWWBBBWBWBBWWWBBBWWBBWBBBWW
BBWBBWWWBWWBBWWBWWWWBWWWBBWW
WBWWBBWWBWWBBBWWBWWBBBBWWBWW
WWBBBWBWBWWWBWWWBBWBWWBWWWBW
WWWBBWWBWWWBBWWBBBWWWBWBBWWB
WBBWWWBWWWBBWWBBWWBBWWWBBWWB
BWBBWWBBBWWBBWBBWWBBWWWBWWBB
WWBBBBBW ...
```

Readers should complete a table like Table 4.4 and check their results against those in Table 4.5. Then they should complete the chi-squared tests and see what their conclusions are.

Remember that to find $H = \sum_i p_i \log_2 (1/p_i)$ for any n-gram, we can use the computational formula

$$\hat{H} = \log_2 \hat{n} - (\frac{1}{\hat{n}}) \sum_i \hat{n}_i \log_2 \hat{n}_i$$

Also recall that the Nth order estimate of H is

$$\hat{H}_N = \hat{H} \text{ (n-gram)} - \hat{H} \text{ (}N - 1 \text{ gram)}$$

We can examine H_N as a function of the order of estimation N.

4.13 Use of the LRX^2 to Determine Order

We need to introduce some new notation. Suppose there are s states and we are interested in testing the hypothesis that the system is a Markov process of order r versus the hypothesis that the system is Markov of order $r - 1$. Let $i_1 i_2 i_3 ... i_r$ be a particular sequence of states and $n_{i_1 i_2 ... i_r}$ be how often that particular sequence of states occurred. Let

$$n_{+i_2 i_3 ... i_r} = \sum_{i_1} n_{i_1 i_2 i_3 ... i_r} \quad ,$$

$$n_{i_1 i_2 i_3 ... i_{r-1}+} = \sum_{i_r} n_{i_1 i_2 i_3 ... i_r} \quad , \text{ and}$$

$$N_r = \sum_{i_1, i_2, ..., i_r} n_{i_1 i_2 ... i_r} = NOBS - (r - 1)$$

For example,

if $n_{A_1 A_1 A_1} = 3$ and $n_{A_1 A_1 A_2} = 4$, then $N_{A_1 A_{1+}} = 7$

if $n_{A_1 A_1 A_2} = 3$ and $n_{A_2 A_1 A_2} = 2$ then $n_{+A_1 A_2} = 5$.

Table 4.5

Quartogram	n_i	$n_i \log n_i$		Trigram	n_i	$n_i \log n_i$
WWWW	3	4.755				
				WWW	19	80.711
WWWB	16	64.000				
WWBW	15	58.603				
				WWB	39	206.131
WWBB	24	110.039				
WBWW	16	64.000				
				WBW	20	86.439
WBWB	4	8.000				
WBBW	22	98.108				
				WBB	31	153.580
WBBB	9	28.529				
BWWW	16	64.000				
				BWW	39	206.131
BWWB	23	104.042				
BWBW	4	8.000				
				BWB	11	38.054
BWBB	7	19.651				
BBWW	23	104.042				
				BBW	30	147.207
BBWB	7	19.651				
BBBW	9	28.529				
				BBB	11	38.054
BBBB	2	2.000				
Sum	200	785.949		Sum	200	956.307

$\hat{H}$(quartogram)

$\quad = 7.644 - \dfrac{1}{200} \times 785.949$

$\quad = 3.71$

$\hat{H}_4 = \hat{H}$(quartogram) $- \hat{H}$(trigram)

$\quad = 3.714 - 2.862 = .852$

$\hat{H}$(trigram)

$\quad = 7.644 - \dfrac{1}{200} \times 956.307$

$\quad = 2.862$

$\hat{H}_3 = \hat{H}$(trigram) $- \hat{H}$(digram)

$\quad = 2.862 - 1.989 = .873$

$T_3 = \hat{H}_3 - \hat{H}_4$

$\quad = .873 - .852$

$X^2 = 5.82$

$T_2 = \hat{H}_2 - \hat{H}_3$

$\quad = .994 - .873$

$X^2 = 33.55$

Table 4.5 (continued)

Digram	n_i	$n_i \log n_i$		Symbol	n_i	$n_i \log n_i$
WW	58	339.763				
				W	100	737.732
WB	51	289.294				
BW	50	289.193				
				B	91	592.209
BB	41	219.660				
Sum	200	1130.910		Sum	200	1329.941

$\hat{H}(\text{digram})$

$= 7.644 - \dfrac{1}{200} \times 1130.910$

$= 1.989$

$\hat{H}_i = \log n - \dfrac{1}{n} n_i \log n_i$

$= \log 200 - \dfrac{1}{200} \times 1329.941$

$= 7.644 - 6.649 = .995$

$\hat{H}_2 = \hat{H}(\text{digram}) - \hat{H}_1$

$= 1.989 - .995 = .994$

$\hat{T}_1 = \hat{H}_1 - \hat{H}_2$

$= .995 - .994$

$X^2 = .28$

To test the hypothesis that the chain is of order $r - 1$,

$$LRX^2 = 2 \sum_{i_1, i_2, \dots, i_r = 1} (n_{i_1 i_2 \dots i_r}) \log_e \frac{N_r(n_{i_1 i_2 \dots i_r})}{(n_{i_1 i_2 \dots i_{r-1}+})(n_{+i_2 i_3 \dots i_r})}$$

which is distributed asymptotically as chi square with $df = s^{r-1} (s-1)^2$.
See the appendix for a derivation of the degrees of freedom.
   For example, for $r = 3$.

$$LRX^2 = 2 \sum_{i,j,k=1}^{s} n_{ijk} \log_e \frac{N_r n_{ijk}}{(n_{ij+})(n_{+jk})} \quad ,$$

when

$$n_{ij+} = \sum_{k} n_{ijk} \; ; \; n_{+jk} = \sum_{i} n_{ijk}$$

To see why this equation is the same as Equation (4.3),

$$LRX^2 = 2 \sum_{i} (n_{observed})_i \log_e (\hat{p}_{observed})_i / (\hat{p}_{expected})_i \quad ,$$

simply note that in any table there is the following row:

		consequent, i_r				
		$i_r = 1$	$i_r = 2$	...	$i_r = s$	marginal
antecedent	$i_1, i_2, \dots, i_{r-1}$					$n_{i_1 i_2 \dots i_{r-1}+} =$ sum

Summing over the consequent gives $n_{i_1 i_2 \dots i_{r-1}+}$, the row total. Hence

$$(\hat{p}_{observed})_i = n_{i_1 i_2 \dots i_r} / n_{i_1 i_2 \dots i_{r-1}+} \quad .$$

Similarly we can show that

$$(\hat{p}_{expected})_i = (n_{+i_2...i_r}) / N_r \; .$$

Because this is what we would expect under the null hypothesis that the Markov chain is of order $r - 1$, in which case the rth remotest antecedent makes no difference, we can sum over it, which gives us $n_{+i_2 i_3...i_r}$. Recall that $N_r =$ # of r-grams.

4.14 Order for Specific Sequences

Figure 4.1 illustrates the idea of determining order hierarchically.

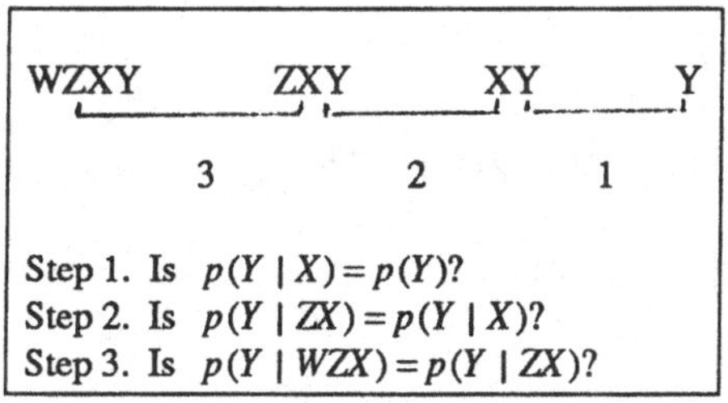

Figure 4.1 Order for specific sequences.

In step 1 we compare the conditional probability $p(Y \mid X)$ to the unconditional probability $p(Y)$. Similarly, in step 2 we compare the two conditional probabilities $p(Y \mid ZX)$ and $p(Y \mid X)$, and so on. The z-score can be computed at each step. For example, in step 2

$$Z = \frac{p(Y \mid ZX) - p(Y \mid X)}{\sqrt{\dfrac{p(Y \mid X)(1 - p(Y \mid X))(1 - p(ZX))}{(n - 2) \, p(ZX)}}}$$

where $p(ZX) = n(ZX) / (n - 1)$, where $n =$ number of observations. In this manner we recommend generalizing the concept of order for specific sequences.

Appendix 4.1

Degrees of Freedom for Order of the Markov Chain

To show that the degrees of freedom for the chi-squared test that a Markov chain with s codes is of order r and not of order $r - 1$, compute the degrees of freedom for one table. When $r = 0$ there are s cells $A_1, A_2, \ldots, A_s$. Since the total is constrained the degrees of freedom are $s - 1$. When $r = 1$ there are s^2 cells $A_1, \ldots, A_s$ by $A_1, \ldots, A_s$, with s margins constrained, so the degrees of freedom are $s^2 - s$. In general, for order r, $df = s^{r+1} - s^r$.

Next we test the hypothesis that the Markov chain is of order r versus that it is of order $r - 1$. The degrees of freedom for the test of these two nested hypotheses is the difference in the number of parameters in the two models:

$$df = s^{r+1} - s^r - (s^r - s^{r-1}) = s^{r+1} - 2s^r + s^{r-1}$$

$$= s^{r-1} (s^2 - 2s + 1) = s^{r-1} (s - 1)^2$$

Appendix 4.2

LRX^2

Assume that we have a random variable $X_1, X_2, \ldots, X_N$, with N fixed, that is, obtained by individual trials with k possible outcomes or categories for each trial. Let $Y_i =$ the number of trials with outcome i. The joint distribution of the Y_i is the multinomial:

$$p[Y_i = y_i \mid i = 1, \ldots, k] = \frac{N!}{y_1! \, y_2! \ldots y_k!} \, p_1^{y_1} \, p_2^{y_2} \ldots p_k^{y_k}$$

where the p's are the probabilities of each category (they sum to 1).

Using likelihood ratio tests, if one model is hierarchically related to another, one with df more parameters than the other, and $p_1(X)$ is the bigger model, and $p_0(X)$ is the smaller (more simplified in the sense that it has *df* fewer parameters), then, asymptotically

$$2\log_e (p_1/p_0) \sim X^2_{df} .$$

This can be employed immediately by writing two models for a multinomial:

Model A: expected counts $A_1, \ldots, A_k$ (these are maximum likelihood estimates)

Model B: expected counts are $B_1, \ldots, B_k$ which are also MLE's under p_0; (this is our null hypothesis, for example). Note that we will let $p_0 =$ Model B and $p_1 =$ Model A.

The likelihood ratios can be formed by dividing

$$p_A(X) = \frac{N!}{X_1!...X_k!} (A_1/N)^{X_1} (A_2/N)^{X_2} ...(A_k/N)^{X_k} \quad \text{by}$$

$$p_B(X) = \frac{N!}{X_1!...X_k!} (B_1/N)^{X_1} (B_2/N)^{X_2} ...(B_k/N)^{X_k}$$

to obtain

$$p_A / p_B = (A_1 / B_1)^{X_1} (A_2 / B_2)^{X_2} ...(A_k / B_k)^{X_k} \quad \text{and}$$

$$2\log (p_A / p_B) = 2\sum X_i \log (A_i / B_i) .$$

If we now let A be the saturated model, i.e., the data itself, $A_i = X_i = O_i$, the observed count. If we let B_i be the expected counts *under some model*, E_i, then we obtain the result that

$$LRX^2 = G^2 = 2\sum O_i \log (O_i / E_i) \sim X^2_{df} ,$$

asymptotically. G^2 is another symbol in common use for the LRX^2.

5

STATIONARITY OF THE MARKOV CHAIN

Our plan in sequential analysis is to compare the likelihood of the occurrence of a sequence for two groups. For example, do fathers smile as often in response to their infants' smiles as mothers do? Obviously, to answer such questions we will have to estimate parameters of sequential connection from our samples and will want to believe that these estimates are, in some senses, "stable". What is meant by "stable?"

Stationarity is a concern with the stability of our parameters of sequential connection over *time*. Homogeneity is a concern with the stability of our parameters of sequential connection across *subjects*. Recall that we only *estimate* these parameters from our sample, so we need tests of statistical significance to decide whether, given our data, the parameters are likely to differ, say from the first to the second half of the interaction.

The reader should consider the assessment of stationarity as a decision that is not automatically made by one significance test. Unfortunately, this has come to be the way stationarity is usually viewed – the test for stationarity is seen only as a precaution that must be taken for our analyses to be valid. We suggest two things: First, the decision about stationarity need not be automatic. It can depend on the researcher's judgment to a much greater degree than has been presumed in the past. Such a decision is much like deciding on an adequate reliability index for a measure; we suggest that the spirit of Cronbach et al.'s (1972) generalizability conceptions can be applied to a discussion of stationarity.

We also want to convey the idea that tests for stationarity can be viewed as an opportunity for the reader to understand the data. For example, Gottman (1979a) discovered that problem decision making for married couples has three distinct phases. In the first phase the couple engages in the task of building an agenda. They get the issues out as each partner sees them, and explore them. In the second phase, they argue for their own position and explore areas of agreement and disagreement. In the third phase, the task is compromise. Our sequences will be different for these three phases and we therefore will not have stationary data.

In the Ekman, Friesen, and Simons (1985) startle experiment, non-stationarity was reflected by the fact that people's facial behavior can be divided into two segments: the relatively stereotyped startle reflex and the more individual emotional reaction to having been startled. So obviously it can be fascinating to uncover shifts in sequential structure.

On the other hand, sometimes the search for sequential structure will not be so fine-grained, and readers won't care to look at things so closely. We suggest they can look at their data with the eye, a magnifying glass, or a microscope, depending on what they're looking for. There is not one right answer about stationarity in the data.

Data are considered stationary if it does not matter when in historical time we started collecting the data. Operationally with one dyad's data, for example, this means that one chunk of the interaction is like another in terms of transition probabilities. This chapter reviews the standard omnibus tests that are used to make the yes-no decision about stationarity. Then we would like to expand the reader's consciousness about the issue.

Our suggestions about stationarity are similar to those currently in the literature. Hewes (1980, p. 409) suggested constructing mathematical models to describe orderly departures from stationarity. Jackson and O'Keefe's (1982) critique of Ellis's (1979) analysis of his data proposed that researchers "compute not only significance tests for nonstationarity, but also estimates of the size of the effect" so that they may judge if departures from stationarity are trivial. Jackson and O'Keefe also proposed examining precisely which transition probabilities change across segments of the data. Such sensible recommendations all address the problems created by applying statistical tests automatically.

5.1 One Dyad

Suppose that the dyad has two states, A and B, and that we divide our data into two segments (see Table 5.1). In Table 5.1 we have computed two transition frequency matrices, one for the first half of the data and another for the second half of the data. We can also compute the overall pooled transition frequency matrix. We seek a test for stationarity that will tell us if the transition frequency matrices for the two segments are different from the pooled matrix, which has ignored time altogether.

Table 5.1

| Segment | Antecedent | Transition frequencies | | Transition probabilities | |
| | | Consequent | | Consequent | |
		A	B	A	B
1	A	10	20	.33	.67
	B	20	10	.67	.33
2	A	15	5	.75	.25
	B	15	25	.38	.62
Pooled	A	25	25	.50	.50
	B	35	35	.50	.50

5.2 Omnibus Test for Stationarity

The likelihood ratio chi-squared test for stationarity is similar to the test for order described in Chapter 4. To test the hypothesis that the Markov chain is of order r against the null hypothesis that the claim is of order $r - 1$, we collapsed the rth order timetable over one dimension to get expected values under the null hypothesis. Then we compared observed values to expected values to compute LRX^2.

In testing for stationarity a similar process is used. We want to see if the data divided into T segments are similar to the data that would be obtained if no such divisions of the data were made. The value for T is usually set at $T = 2$ unless there is a theoretical reason to use another way of chunking the data stream into segments. The expected transition frequencies and probabilities under the null hypothesis are those that collapse the data over the T segments. If the chain is of order r, then let $n_{i_1 i_2 ... i_r}(t)$ denote the rth order transition frequency for a particular sequence $i_1 i_2 ... i_r$ for segment $t (1 \leq t \leq T)$. Let $p_{i_1 i_2 ... i_r}(t)$ denote the transition probability of that sequence for segment t, and denote $p_{i_1 i_2 ... i_r}$ the expected transition probability under the null hypothesis that the data are stationary. Then an omnibus test of stationarity is

$$LRX^2 = 2\sum_{t=1}^{T} \sum_{i_1,i_2,\ldots,i_r=1} n_{i_1 i_2 \ldots i_r}(t) \log_e \frac{\hat{p}_{i_1 i_2 \ldots i_r}(t)}{\hat{p}_{i_1 i_2 \ldots i_r}}$$

which is asymptotically distributed as chi square with degrees of freedom equal to $(T-1)(s)^r(s-1)$.

For a first-order Markov chain (a two-dimensional table), the omnibus test of stationarity is

$$2\sum_{t=1}^{T} \sum_{j=1}^{s} \sum_{i=1}^{s} n_{ij}(t) \log_e \frac{\hat{p}_{ij}(t)}{\hat{p}_{ij}} \sim X^2_{s(s-1)(T-1)}$$

asymptotically, for large n. In Table 5.1, $S = 2$ (two states) and $T = 2$ (two segments). The $\hat{p}_{ij}$ are the pooled transition probabilities, $\hat{p}_{ij}(1)$ are the transition probabilities for segment 1, $\hat{p}_{ij}(2)$ are the transition probabilities for segment 2, and $n_{ij}(t)$ are the respective transition frequencies for each segment. Computation gives $G^2 = 7.25$, which is compared to $X^2_{2(2-1)(2-1)} = X^2_2$. These data would probably be judged nonstationary.

5.3 Other Reasons for Dividing the Data into Segments

We may wish to distinguish between segments of an interaction even if the transition probabilities are stable. The frequencies of our codes may vary from segment to segment. In Table 5.1 the test that the frequencies of codes A and B are stable would result in a X^2 of $3.43(df = 1)$, which may lead us to suspect whether our two segments are not identical from an interactional standpoint. In marital interaction in which the couple is resolving an issue in their marriage, Gottman (1979a) found that the beginning third agenda building phase is more likely to contain expressions of feelings about a problem, the middle third arguing phase is more likely to contain disagreement, and the final third negotiation phase is more likely to contain proposals for solutions to the issues than the other phases, respectively. This suggests that different social processes are probably going on during different parts of a discussion.

Gottman recommended using time-series graphs for an initial assessment of whether the data are the same in different segments of an

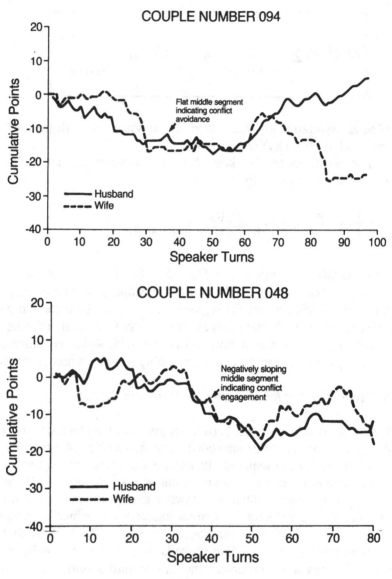

Figure 5.1. Point graphs of two married couples showing different patterns in the middle arguing phase of the discussion.

interaction. The y-axis is a measure of the cumulative number of (positive minus negative) interactions, called "cumulative points." The point graphs in Figure 5.1 represent the problem-solving interaction of two couples who are very different in the way they handled the middle

Table 5.2

	Segment	
	1	2
1	$n_{ij}(1,1)$	$n_{ij}(1,2)$
2	$n_{ij}(2,1)$	$n_{ij}(2,2)$
.		
.		
.		
s	$n_{ij}(s,1)$	$n_{ij}(s,2)$

s = # of subjects

arguing phase. One couple avoided any conflict or disagreement (flat middle) while the other couple engaged in conflict (negatively sloping middle). More detailed sequential analysis of the data for the two couples corroborated this conflict-avoiding versus conflict-engaging interpretation of the point graph. We would not want to consider the whole conversation as uniform if we have such strong reasons to suspect that different behaviors during different segments of the discussion are so revealing of different styles of interaction.

A careful examination of the sequences of couple #94 showed that they indeed did avoid conflict; during the arguing phase they had no disagreement chains (Husband Disagreement → Wife Disagreement, and conversely), but that they had both disagreement, cross-complaining, and counterproposal chains in the final third, or negotiation phase ($N = 7$). On the other hand, couple #48 who indeed engaged in conflict, had two disagreement chains in the arguing phase; however, they had only one cross-complaining chain in the negotiation phase and no arguing or counterproposal chains in the negotiation phase. This suggests the hypothesis that avoiding conflict in the arguing phase leads to poor negotiation. If this can be corroborated in subsequent analysis (we will have to pool subjects on the basis of similar point graphs, assuming homogeneity tests out), it will have been important to break the discussion into three segments.

The preceding paragraphs have been designed to convince readers to approach the issues of stationarity and homogeneity in creative ways, which can help them discover truths about the data.

5.4 An Alternative to the Omnibus Test

For each transition from state i to state j we can form the analysis of variance table in Table 5.2. Usually we are interested in a *specific* set of sequences in our data, and not all possible sequences. Our data may well be homogeneous and stationary regarding the sequences of interest but not with respect to other sequences. Using Table 5.2 we can compute a statistic analogous to Cronbach's alpha, which gives us the amount of variance due to subjects compared to the amount of variance due to segments. If the alpha is high (say, .70 or above) we may decide that the data for this transition (i to j) are stationary regarding this set of subjects. If the frequency of the antecedent varies greatly from subject to subject, it can be used as a covariate.

6

HOMOGENEITY

Before pooling data across subjects to create one overall timetable for all subjects, we need to determine if the data from different subjects have essentially the same sequential structure. In a contextual design we may wish to assess homogeneity separately within each cell of the design. Suppose we have NCODES = the number of codes in our observational category system.

The assessment of homogeneity is an opportunity for discovery because it can lead us to block subjects into homogeneous groups. Suppose we have determined that the Markov chain is of order r.

To test for homogeneity we proceed as we do when testing for stationarity. Suppose we have s subjects. The Anderson-Goodman test statistic is

$$LRX^2 = 2 \sum_{j=1}^{s} \sum_{i_1,i_2,\ldots,i_r=1}^{\text{NCODES}} n_{i_1 i_2 \ldots i_r}(i) \frac{\hat{p}_{i_1 i_2 \ldots i_r}(j)}{\hat{p}^{*}_{i_1 i_2 \ldots i_r}} \tag{6.1}$$

where $\hat{p}^{*}$ are computed across all subjects, pooled. Recall that r is the order of the chain, that $i_1, i_2 \ldots i_r$ denotes a particular sequence, and $\hat{p}_{i_1,i_2 \ldots i_r}(j)$ is the transition probability of that sequence for subject j. The degrees of freedom are $df = (s - 1)(\text{NCODES})^r(\text{NCODES} - 1)$. We deal with empty cells by subtracting the number of empty cells from the degrees of freedom.

The following are examples from published studies that have examined order, stationarity, or homogeneity.

6.1 Hirokawa (1980)

Twenty-three four-member groups were formed to solve the NASA moon shot problem. The problem had a correct solution, so it was possible to divide groups on an effectiveness dimension. The four most and least effective groups were coded (there were 1,667 utterances). When data were collapsed over time there were no significant differences between types of groups.

Order. The Anderson-Goodman test for order showed that the data fit a first-order model (see Table 6.1).

Stationarity. Each of the eight discussions was divided into two equal segments. The Anderson-Goodman test showed that the data were stationary (see Table 6.2).

Homogeneity. The Anderson-Goodman tests for homogeneity within effective and ineffective groups showed that the data were homogeneous within groups (see Table 6.3).

Group comparisons. Effective and ineffective groups were compared in Table 6.4. The social interactions in the two groups were significantly different, that is, not homogeneous.

Lag sequential analyses. Only the four effective groups displayed a two-behavior pattern of procedural statement providing direction followed by procedural agreement. They wrote:

> This suggests that group members in the effective groups consistently interacted with each other on procedural concerns (e.g., criteria for making a decision), while group members in the ineffective groups apparently spent much of their time interacting directly on task relevant matters. (p. 319)

6.2 Sillars (1980)

The hypothesis tested in this study of college roommate interaction was based on the so-called fundamental attribution error (overestimating the role of dispositional vs. situational causes of behavior).

Table 6.1. Likelihood ratio values for the Anderson-Goodman test for order effects (from Hirokawa, 1980).

Discussion	Hypothesis	LR	df	p
Effective group 1	0-Order vs. 1st-Order	113.84	16	.001
	1st-Order vs. 2nd-Order	50.28	80	ns
Effective group 2	0-Order vs. 1st-Order	57.04	16	.001
	1st-Order vs. 2nd-Order	33.83	80	ns
Effective group 3	0-Order vs. 1st-Order	68.55	16	.001
	1st-Order vs. 2nd-Order	32.15	80	ns
Effective group 4	0-Order vs. 1st-Order	107.67	16	.001
	1st-Order vs. 2nd-Order	53.17	80	ns
Ineffective group 5	0-Order vs. 1st-Order	131.26	16	.001
	1st-Order vs. 2nd-Order	53.26	80	ns
Ineffective group 6	0-Order vs. 1st-Order	33.27	16	.001
	1st-Order vs. 2nd-Order	31.47	80	ns
Ineffective group 7	0-Order vs. 1st-Order	106.49	16	.001
	1st-Order vs. 2nd-Order	61.83	80	ns
Ineffective group 8	0-Order vs. 1st-Order	46.61	16	.001
	1st-Order vs. 2nd-Order	12.78	80	ns

Table 6.2. Likelihood ratio values for the Anderson-Goodman tests for stationarity (from Hirokawa, 1980).

Discussion	Segment	LR	df	p
Effective group 1	2 segments	24.67	20	ns
Effective group 2	2 segments	17.09	20	ns
Effective group 3	2 segments	8.30	20	ns
Effective group 4	2 segments	23.24	20	ns
Ineffective group 5	2 segments	20.43	20	ns
Ineffective group 6	2 segments	14.95	20	ns
Ineffective group 7	2 segments	19.09	20	ns
Ineffective group 8	2 segments	10.02	20	ns

Table 6.3. Likelihood ratio values for the Anderson-Goodman tests
for homogeneity (from Hirokawa, 1980).

Hypothesis	LR	df	p
Interaction within effective groups (1,2,3,4) are not homogeneous with each other	72.30	60	ns
Interaction within ineffective groups (5,6,7,8) are not homogeneous with each other	59.49	60	ns

Table 6.4. Likelihood ratio values for the Anderson-Goodman test
for homogeneity between effective and ineffective discussions
(from Hirokawa, 1980).

Hypothesis	LR	df	p
Interaction within effective groups (1,2,3,4) are not homogeneous with interaction within ineffective groups (5,6,7,8)	78.36	20	.001

Forty-six pairs of same-sex college-age roommates engaged in a
conversation about problems they had experienced as roommates. They
also filled out a questionnaire about 16 types of roommate grievances
and they were asked to attribute the cause of the problem to self or roommate, and the perceived stability of the problem.

Tapes were coded using the Raush et al. (1974) system as
avoidance (conflict avoiding), *distributive* (recognizes problems but
negatively evaluates roommate or seeks compliance), and *integrative*
(more positive).

This paper is interesting in that it employed a second-order Markov
model. It also used log-linear analysis (see Chapter 10) which makes it
possible to assess how the timetable changes as a function of contextual
factors. The likelihood ratio chi square was used to assess the statistical
significance of effects. There was a significant interaction of internal-
external, stable-unstable, and the first-order 3×3 timetable (time zero by
time one factors), with $LRX^2 = 9.43$, $df = 4$, $p = .051$. Chapter 10
discusses the assessment of effects like these.

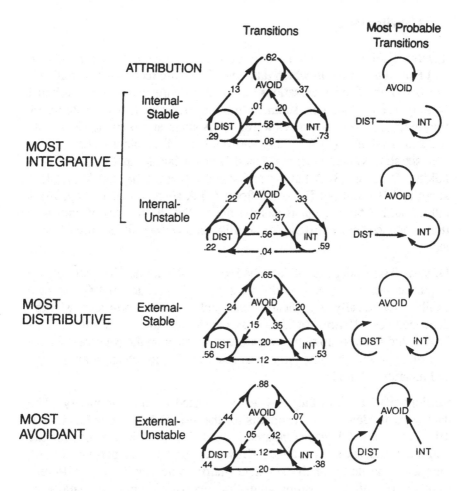

Figure 6.1. First-order transition probabilities of subjects making internal-stable, internal-unstable, external-stable, and external-unstable attributions (from Sillars, 1980).

Figure 6.1 is a brief summary of Sillars's analyses with state transition diagrams. Some conclusions drawn from Figure 6.1 are: (1) There was a greater tendency to follow each type of act with avoidance given an external-unstable attribution; (2) subjects were more likely to escalate conflicts and less likely to deescalate conflict when they made external rather than internal attributions.

6.3 Capella (1980)

Twelve dyadic conversations were observed using the procedure of Jaffe and Feldstein (1970) in which talk and silence are recorded automatically by computer. The conversations lasted 25-30 minutes; the first 5 minutes were ignored. Data were summarized into four-state and six-state transition matrices. The four states were: (1) A silent and B silent; (2) A talking and B talking; (3) A talking and B silent; (4) A silent and B talking. The six state system was: (1) A and B silent but A has the floor; (2) A talking, B silent; (3) A talking and B talking but A has the floor; (4) A silent and B silent but B has the floor; (5) A silent and B talking; (6) A talking and B talking but B has the floor. The 20 minutes of conversation produced four thousand 300-millisecond observations for each subject.

Order. Two independent labs had previously found that talk-silence sequences sampling at two different rates (0.166 and 0.300 seconds) exhibit statistically significant but negligible second-order Markov dependencies. Hence, the data was considered to be first-order. This kind of judgment is up to the researcher to justify on the basis of the content of the research. The job of the researcher is to select an interesting and meaningful model.

Stationarity. Each of the 12 dyads was tested for nonstationarity. The data were divided into 10 segments for the tests and the alpha level set at .01. Time period 1 was compared with 2, 3 with 4, and so on, for the four-state and six-state matrix. Only one of the time periods for the four-state and none for the six-state matrix were significantly different from the overall composition transition matrix. Overall, the data were considered stationary. Once again this decision was a judgment call by the researcher, which is perfectly appropriate.

Homogeneity. Each of the 12 dyads was significantly different from the composition matrix, and the situation was not improved by considering male-male, male-female, and female-female composites separately. The data were not homogeneous.

6.4 Valentine and Fisher (1974)

This study, which presents a creative use of stationarity analysis, was a study of small group interaction.

A 7-category system was employed to the interaction of 6 groups. The talk turn was the unit of analysis on the basis of an analysis with another coding system. The authors hypothesized that groups go through four phases (orientation, conflict, emergence, and reinforcement). The Pearson X^2 test was used to compare phases. A summary description of each of the four phases follows (see Table 6.5).

Orientation. There is a relative absence of others attacking or ignoring an assertion; members tend toward "clarification" as a safe way of disagreeing. Group members are reluctant to take too strong a stand until the group has developed.

Conflict. There is a lot of simple disagreement, confrontation, and pressure to conform. Clarification, neutrality, and safe interaction are unacceptable during this second phase, and confrontation is maximized. Coalitions develop around positions.

Emergence. This is described as a transition phase; it appears to involve the reduction of conflict.

Reinforcement. This stage reflects approaching consensus. There was no persistent response to disagreement and conflict. Here we have an example of nonstationarity leading to an interesting substantive hypothesis about the phases of group interaction.

6.5 Cline (1979)

Nine pairs of unacquainted college students, nine pairs of roommates, and nine couples were asked to discuss the proposition "Child-rearing practices in America should become more permissive." The author was interested in conversation that focused on the relationship (on the developing relationship). Chi-squared tests showed the Markov matrices were not independent. Dyadic statements were rare; people seldom focused on their relationships. The hypothesis that people are more likely to focus on their own relationships in the early stages of acquaintanceship was not supported; however, when a dyadic statement was offered, another dyadic statement was more likely to follow it than expected (p. 67).

Table 6.5(a). Static analysis (all phases)

Phases	Interaction categories						
	1	2	3	4	5	6	7
Orientation	29	15	13	2	8	9	308
Conflict	84	48	34	12	26	14	209
Emergence	50	44	23	5	14	8	416
Reinforcement	15	14	11	8	10	3	245

$n = 1877$

Table 6.5(b). Contiguity analysis (orientation phase)

Antecedent units	Subsequent units						
	1	2	3	4	5	6	7
1	5	2	3	1	6	0	12
2	12	12	0	0	0	0	1
3	0	1	7	0	0	0	5
4	1	0	0	0	0	0	1
5	0	1	0	0	3	1	3
6	0	0	0	0	0	5	4
7	10	10	1	3	0	3	281

$n = 384$

Table 6.5(c). Contiguity analysis (conflict phase)

Antecedent units	Subsequent units						
	1	2	3	4	5	6	7
1	15	11	15	0	13	1	29
2	31	3	0	0	2	0	12
3	12	1	5	1	2	0	13
4	0	1	1	0	1	0	9
5	3	1	5	1	6	0	10
6	2	2	0	0	1	5	4
7	10	30	8	10	0	8	334

$n = 627$

Table 6.5(d). Contiguity analysis (emergence phase)

Antecedent units	Subsequent units						
	1	2	3	4	5	6	7
1	4	7	9	1	9	1	19
2	23	3	0	0	4	0	14
3	4	2	5	0	2	1	9
4	0	1	1	0	0	0	3
5	4	3	4	0	1	0	2
6	0	2	1	0	0	2	3
7	15	27	4	5	0	4	361

$n = 560$

Table 6.5(e). Contiguity analysis (reinforcement phase)

Antecedent units	Subsequent units						
	1	2	3	4	5	6	7
1	1	2	4	1	1	0	6
2	3	0	0	0	1	0	10
3	5	0	1	1	1	0	3
4	0	0	1	0	2	0	5
5	0	0	0	1	3	0	6
6	0	0	0	0	0	2	1
7	6	13	6	6	2	1	211

$n = 306$

6.6 Hawes and Foley (1973)

Sixteen initial medical interviews conducted by four physicians were coded with 13 categories, which produced a three-state model with categories: inhibit, maintain, or facilitate communication. Stationarity analyses were conducted for the 13-category system comparing beginning, middle, and end of the 16 interviews. Although it was not an ideal method, correlation coefficients were computed across categories: Beginning with Middle, $r = .962$; Beginning with End, $r = .959$; Middle with End, $r = .964$. Order was approached using an entropy statistic suggested by Shannon and Weaver (1949) and the conclusion was that the states were not independent. Interviewer effects were compared using the three-category system that collapsed over time. These analyses and the state transition diagrams showed that the interviewers were quite different. Some of their conclusions were: (1) Objective questions and subjective questions were found to be initiator states. They started a cyclic pattern but were not themselves part of the cycle. In the objective question cycle objective information was obtained, extended, clarified, and positively reinforced. In the subjective question cycle subjective information was extended, clarified and structured. (2) Both criticism and generalizing led to inhibitions of communication and back to an objective question cycle.

6.7 Lichtenberg and Hummel (1976)

Six initial therapy interviews, all by famous psychotherapists, four by Albert Ellis and two by Carl Rogers, were analyzed using a four-category coding system: (a) personal; (b) descriptive; (c) cognitive; and, (d) directive. Two additional categories were "introductory" and "terminal." The data were divided into first and second halves and the chi-squared test resulted in a nonstationarity decision. A randomization test procedure was used rather than the usual chi square, and the first-order Markov property was rejected in favor of an "independence" model for client and therapist responses. As far as can be detected from the coding system Lichtenberg and Hummel employed, these two famous psychotherapists, one noted for being extremely directive (Ellis), the other for being extremely nondirective (Rogers), do psychotherapy independent of their clients, which is quite an interesting result.

6.8 Hawes and Foley (1976)

Seven meetings of three academic committees were videotaped, transcribed, and coded. In general, the data from the three committees were stationary, using the likelihood ratio chi-squared test. Using information theory, they decided that the data fit a first-order Markov process. The data were *not* found to be homogeneous, using a LRX^2 test. The major function of this paper was to demonstrate how these computations were to be performed.

7

EVERYDAY COMPUTATIONS OF STATIONARITY,
ORDER AND HOMOGENEITY

7.1 The Arundale Programs

Arundale (1982) wrote two useful programs for Markov analysis, called SAMPLE and TEST. The program SAMPLE takes as input a sequence of as many as 3,000 observations of up to 15 different discrete states. It outputs transition frequency and transition probability matrices for the input sequence. SAMPLE also tests the stationarity of the sequence (user specifies the number of chunks to break the original data set into) and it also tests order. The program TEST takes as input a set of transition frequency matrices and compares pairs of these matrix sets by specifying one set as the expected values and one set as the observed values. Both Pearson and likelihood ratio statistics are computed. It is possible to average a set of the input matrices before testing, or to test a group of matrices for homogeneity. The Arundale programs are also now available for an IBM-compatible personal computer.

7.2 Example

The data in this chapter come from a study, conducted by Krokoff, Gottman, and Roy (1988), of 120 married couples in a random sample study of blue- and white-collar couples who were either happily or unhappily married. Fifty-two of the home audiotapes of these couples attempting to resolve an existing marital issue were coded using the Couples Interaction Scoring System (see Gottman, 1979a). In the present context we examine only the affect portion of these codes, of which there are six: (1) husband positive (H+); (2) husband neutral (H0); (3) husband negative (H–); (4) wife positive (W+); wife neutral (W0); and, (6) wife negative (W–). We have reason to suspect that in this particular coding system, the neutral and positive codes function in a similar manner (in the context of marital conflict), and we will combine them for some of our analyses.

The following data was input to Arundale's program SAMPLE for Couple #145, an unhappily married white-collar couple (U-WC).

```
2 5 2 4 2 5 2 4 2 4 2 2 5 2 4 2 5 3 4 3 4 1 4 4 1
2 5 2 1 5 4 2 5 1 4 4 2 3 6 6 2 2 6 5 5 5 6 3 6 3
3 6 3 2 5 3 5 2 6 2 2 2 5 2 4 4 5 2 5 2 5 5 4 2 5
5 5 2 5 2 5 2 5 2 2 5 2 5 2 2 5 5 2 2 2 5 2 5 2 5
2 5 2 2 5 2 5 3 5 2 5 5 2 2 5 2 2 5 2 6 2 5 4 2 5
2 4 4 2 1 2 5 2 5 2 5 5 5 3 5 2 5 5 2 5 5 2 5 2 5
2 4 2 4 2 5 4 2 2 5 2 5 5 5 2 5 2 5 5 2 5 2 5 5 2
1 2 5 1 5 1 5 5 4 2 2 2 3 6 3 6 3 6 3 6 3 3 6
6 6 5 1 2 5 2 5 5 2 4 5 5 2 5 5 2 5 2 5 2 5 1 5 4
2 2 5 2 5 2 5 4 1 4 4 2 4 4 2 4 2 3 5 4 2 2 5 2 6
1 4 1 4 5 5 5 4 5 2 4 5 5 2 5 2 2 2 5 2 2 5 2 2 4
2 2 2 6 2 4 3 3 3 3 2 3 6 3 5 3 3 5 3 2 2 2 2 6 3
6 3 6 2 6 1 6 3 3 3 3 5 3 5 2 5 3 3 6 3 1 4 1 5 1
5 6 3 3 6 3 6 3 6 3 6 3 6 3 3 6 3 6 3 2 5 2 5 2 5
2 5 2 2 2 3 5 1 6 1 6 6 6 1 6 6 1 6 1 6 2 5 1 6 6
6 2 2 6 6 6 2 6 2 6 2 5 5 5 5 5 5 6
```

Table 7.1 is a summary of the transition frequency matrix for these data.

7.3 Stationarity

Gottman (1979a) suggested that there are three distinct stages of a marital conflict discussion of an existing area of disagreement: (1) the agenda-building phase, in which the problem is described and defined; (2) the arguing phase, in which areas of agreement and disagreement are elaborated; and, (3) the negotiation phase, in which the goal is to reach a compromise decision. He suggested that the middle arguing phase should have more negative affect than the first and last phases. This is a nonstationary model.

Table 7.2 summarizes the three probability transition matrices of couple #145's discussion.

The overall test for stationarity has 60 degrees of freedom and the LRX^2 was 68.90. Hence, these data must be considered stationary, using the omnibus test. In part this is because the arrays are so sparse. Small N's imply little power and, hence, small statistics. Note, however, that the

Table 7.1. Transition frequency matrix for Couple #145

	time $t+1$					
time t	H+	H0	H–	W+	W0	W–
H+	0	4	0	6	6	7
H0	3	28	5	14	63	10
H–	1	4	12	2	9	20
W+	5	19	2	6	5	0
W0	8	57	8	9	27	3
W–	6	10	21	0	2	10

Table 7.2. Stationarity analysis for Couple #145

	time $t+1$					
time t	H+	H0	H–	W+	W0	W–
H+	.00	.25	.00	.50	.25	.00
H0	.03	.03	.03	.13	.72	.08
H–	.00	.13	.00	.25	.25	**.38**
W+	.17	.67	.08	.00	.08	.00
W0	.03	.76	.09	.09	.00	.03
W–	.00	.43	**.43**	.00	.14	.00
H+	.00	.38	.00	.25	.38	.00
H0	.06	.00	.06	.14	.71	.03
H–	.00	.00	.00	.00	.25	**.75**
W+	.08	.77	.00	.08	.08	.00
W0	.13	.69	.03	.16	.00	.00
W–	.14	.00	**.72**	.06	.14	.00
H+	.08	.00	.00	.17	.17	.58
H0	.00	.00	.10	.14	.48	.29
H–	.05	.15	.00	.00	.25	**.55**
W+	.17	.17	.17	.00	.50	.00
W0	.15	.50	.20	.05	.00	.10
W–	.20	.28	**.52**	.00	.00	.00

conditional probability for negative affect reciprocity increased for the middle arguing phase, as predicted.

7.4 Order

The comparison that the sequence is zero order rather than first order is the independence test. It produced an LRX^2 of 340.03 with 25 degrees of freedom. This means there is sequential dependency in these data. The next test of first versus second order produced an LRX^2 of 162.41 with 150 degrees of freedom, not significant. Thus, we get no improvement in model fitting in going from first to second order.

7.5 Homogeneity

To illustrate the homogeneity analyses, we compared an unhappily married blue-collar couple (couple #20) with an unhappily married white-collar couple (couple #145). With 21 degrees of freedom, the LRX^2 was 256.62, a highly significant difference. Table 7.3 summarizes the transition probability matrices of these two couples. This analysis indicates that the social class variable may be important in this study.

7.6 Homogeneity Analyses as an Opportunity for Discovery

Let us begin with the full 2×2 factorial design (blue collar/white collar by happily/unhappily married). Within each cell there were 13 transition frequency matrices — one transition frequency matrix for each married couple; each matrix was 6×6; the six codes were husband neutral, positive, or negative affect (H0, H+, H–), and wife neutral, positive, or negative affect (W0, W+, W–). As we began analyzing these data we discovered that they were not homogeneous. Table 7.4 illustrates the lack of homogeneity we discovered in these data. Within each cell of the contextual design the likelihood ratio chi-squared test for homogeneity produced evidence for inhomogeneity.

   To deal with these inhomogeneity problems, Krokoff suggested there were two kinds of married couples in each cell of the design: those who avoid conflict, and those who engage in conflict while discussing an existing marital issue. This dimension of avoiding or engaging in conflict was related to a whole cluster of variables we measured about the couples' philosophy of marriage. Conflict avoiding couples did not have a ''companionate'' philosophy of marriage. They did not view

Table 7.3. Homogeneity analysis for unhappily married
white- and blue-collar couples

time *t*	time *t*+1					
	H+	H0	H–	W+	W0	W–
Couple #145						
(unhappy white-collar)	.00	.17	.00	.26	.26	.30
	.03	.00	.05	.15	.66	.11
	.03	.11	.00	.06	.25	**.56**
	.16	.61	.07	.00	.16	.00
	.09	.67	.09	.11	.00	.04
	.15	.26	**.54**	.00	.05	.00
Couple #20						
(unhappy blue-collar)	.00	.00	.00	.00	.00	.10
	.00	.00	.11	.07	.55	.27
	.00	.03	.00	.05	.43	**.48**
	.00	.00	.67	.00	.17	.17
	.00	.34	.57	.00	.00	.09
	.01	.17	**.76**	.03	.04	.00

Table 7.4. Homogeneity analyses for 2 × 2 design

Group	LRX^2	df	Z^*
Unhappy white-collar	688.8	360	10.3
Unhappy blue-collar	669.8	360	9.8
Happy white-collar	770.1	360	12.4
Happy blue-collar	757.7	360	12.1

* Z is computed as $\sqrt{2X^2} - \sqrt{2df - 1}$

Table 7.5. Homogeneity test for eight cell design

Group	LRX^2	df	Z
Unhappy white-collar conflict engagers	360.2	180	7.9
Unhappy white-collar conflict avoiders	206.4	150	3.0
Happy white-collar conflict engagers	362.6	180	8.0
Happy white-collar conflict avoiders	326.5	150	8.3
Unhappy blue-collar conflict engagers	216.3	150	3.5
Unhappy blue-collar conflict avoiders	378.3	180	8.6
Happy blue-collar conflict engagers	308.5	180	5.9
Happy blue-collar conflict avoiders	371.4	150	10.0

companionship or intimacy as a major goal of the marriage. Instead, they tended to have a traditional role-related view of marriage and family. They tended not to view each other as resources for getting over bad moods, to discuss their day at the end of each day, or to try to come to consensus about areas of disagreement.

Krokoff had designed a questionnaire called "Philosophies" to tap this dimension. The items on it included "I think people should work out their personal problems alone" and "I don't like to rock the boat by bringing up problems with my spouse."

We split the couples within each cell of our 2×2 design at the median of the couples' average score on philosophies (husband and wife scores correlated significantly). We then performed our homogeneity analyses within each cell. As can be seen from Table 7.5, these chi squares are much smaller (the expected value of chi square is df). The average z-score value is now 6.9, reduced from 11.2.

To explore whether there are larger contextual effects, compared to any inhomogeneity that may exist, the Arundale programs were employed to compare across cells. Comparing unhappy white-collar conflict engagers with unhappy blue-collar conflict engagers yielded an LRX^2 of 678.5 with $df = 360$. A comparison of unhappy white-collar conflict engagers with happy white-collar conflict engagers yielded an LRX^2 of 825.2 with $df = 390$. In this manner we can get some initial assessment of the relative contribution of our three contextual factors (blue/white; happy/unhappy; avoid/engage). Of course, the most interesting results are to be found in an analysis of particular cells of the timetable that have theoretical interest (e.g., the reciprocity of negative and positive affect).

7.7 Setting Alpha Levels

A number of writers have noted that the omnibus tests for homogeneity and stationarity depend on the amount of data one has collected.

Capella (1980) discussed the choice of the alpha level for the omnibus test of stationarity. He wrote:

> The choice of confidence levels at 0.01 deserves some discussion since if α were set at 0.05, four of the five values for the six state would be considered statistically significant. The problem is this: with huge amounts of data, small discrepancies between probabilities in the observed and expected matrices are magnified by the expected row frequency, which is usually large. The numerator of the X^2 stationarity statistic is the square of an observed minus expected probability times the row frequency of the expected matrix. Since the expected matrix for any dyad consists of 400 observations, any row frequency could easily be in the 50-100 range as a multiplying factor of the discrepancy. Thus, what appear to the researcher's eye as negligible discrepancies are greatly magnified because of the large data base upon which the statistics are calculated. The problem is further compounded because the degrees of freedom for the X^2 test takes into account *only* the number of independent *cells* in the transition matrix, and not the number of observations upon which they are based. (p. 133)

Capella thus set alpha as small as possible while maintaining reasonably high power. This discussion argues for using the omnibus tests not in an automatic way but rather as a guide for data exploration. In most applications this will not be an issue and our problem will be that the tests are relatively insensitive. However, if we have conducted a large study relative to the number of coding categories, or if we sample often (e.g., as in the Jaffe and Feldstein [1970] paradigm), then the omnibus tests will be too sensitive. For example, Capella (1980) used an automated method for coding the state of two speakers (using talk and silence codes detected by directional microphones and corrected by specially designed software). Because of the enormous number of observations for each conversation, it was nearly impossible to obtain homogeneity without setting the alpha levels for the Anderson-Goodman test lower. This is a perfectly reasonable procedure in our view. The tests should be used not automatically but in a reasoned fashion, as part of selecting the best-fitting, most parsimonious, most elegant, and theoretically the most interesting model.

8

SAMPLING DISTRIBUTIONS

This chapter reviews a few fundamental distributions that will be useful. Many of the z-scores that have been proposed differ in their assumptions about the underlying sampling distribution. Much of this chapter will be familiar to readers who have taken a beginning course in probability and statistics. Nonetheless, we feel we should go over a few basic concepts before proceeding. Our review will be restricted to the counting distributions, primarily the binomial, Poisson, negative binomial, hypergeometric, multinomial, and product-multinomial. We will not discuss the waiting time distributions (geometric, gamma, and negative hypergeometric).

8.1 Binomial Distribution

Suppose N independent events occur, that each event is either event E, which we will call a success (S), or not event E, which we will call a failure (F), that p is the probability of a success, and that $q = 1 - p$ is the probability of a failure. Then if we want to compute the probability of x successes in N trials it is

$$P[X = x] = B(N, p) = \binom{N}{x} p^x q^{N-x}$$

$$\binom{N}{x} = \frac{N!}{x! \, (N-x)!}$$

This latter quantity is the number of ways of choosing x things from a group of N. It is $\dfrac{N(N-1)...(N-x+1)}{x(x-1)...1}$. The numerator represents the total number of ways of selecting x things from N (for the first choice there are N possibilities, for the second choice there are $N-1$

possibilities, and so on). The denominator represents the number of permutations (rearrangements of the x things selected).

The quantity

$$p^x q^{N-x}$$

represents the probability of any string with x successes and $N - x$ failures.

The binomial distribution is also the expansion of the binomial $(p + q)^N$. The $(k + 1)^{th}$ term of the expansion is

$$\binom{N}{k} p^k q^{N-k}$$

The binomial distribution has a long history. It was derived by James Bernoulli in his *Ars Conjectandi* in 1713, although binomial coefficients occur in earlier works (e.g., Pascal).

The mean and variance of the binomial distribution are

$$\mu = Np$$

$$\sigma^2 = Npq$$

To see this, let

$$x = x_1 + x_2 + ... + x_N$$

where x_1 is 1 or 0 depending on whether the outcome of the ith trial was a success or a failure, respectively. Then, since the trials are independent,

$$E(x) = \mu = E(x_1 + ... + x_N) = E(x_1) + ... + E(x_N)$$

$$\mu = Np$$

$$var(x) = var(x_1) + ... + var(x_N)$$

$$= N(E(x_i^2) - (E(x_i))^2)$$

$$= N(p - p^2) = Np(1-p) = Npq \ .$$

The central limit theorem guarantees that the sum $x_1 + x_2 + ... + x_N$ is asymptotically normal. Asymptote is nearly reached when $N > 100$. Really, it is never reached, but when $N > 100$, the large sample approximation is good. When N is about 25 (and < 25), a continuity correction of $\frac{1}{2}$ is necessary. For example, if $N = 30$ and $p = .4$, we can estimate $p\{15 \le x \le 20\}$ by $p\{14\frac{1}{2} \le \text{Normal} \le 20\frac{1}{2}\}$ in which the mean and variance of the normal distribution are $\mu = Np = 12$ and $\sigma = \sqrt{Npq} = 2.68$. The normal estimate of $p\{15 \le x \le 20\}$ is the area under the normal distribution with $\mu = 12$ and $\sigma = 2.68$ between $14\frac{1}{2}$ and $20\frac{1}{2}$.

8.2 Poisson Distribution

The Poisson distribution (published in 1837) is similar to the binomial with N large and p small. It describes events occurring in continuous time with three properties:

1. Events occur with a constant average rate, λ events per unit time;

2. no matter how close and how small, separate time intervals are independent; and

3. events cannot occur simultaneously.

If we denote by $X(t)$ the number of events in an interval from 0 to t, then

$$P(X(t) = x) = (\lambda t)^x \, e^{-\lambda t} / x!$$

The average number of events from 0 to t is λt.

The Poisson distribution is a limit of binomials if the time interval is chopped up into small intervals of size Δt. If we have N of these small intervals, the $\Delta t = t/N$. In each interval the probability of one success is approximately $\lambda \Delta t = \lambda t/N$. Hence, the probability of no successes is approximately $1 - \lambda t/N$, plus a term of order t/N which goes to zero as Δt gets small. Thus the probability of x success is

$$\binom{N}{x} (\lambda t/N)^x (1 - \lambda t/N)^{N-x}$$

As $n \to \infty$ this approaches

$$\frac{(\lambda t)^x}{x!} e^{-\lambda t}$$

Both the mean and variance of the Poisson distribution are λ.

An important property of the Poisson distribution is that if we have two Poisson variables, the "conditional" distribution of each variable, conditional on a given sum of the two variables is a binomial distribution. This result can be generalized to n Poisson variables. Also, the sum of a set of Poisson variables is also distributed as Poisson. The conditional distribution of the variables in the sum is multinomial.

8.3 Negative Binomial

We may be interested in the number of trials necessary to obtain r occurrences of the event, which has a constant probability, p, of occurring at each trial. Let Y_r be the number of failures before the rth success. Then

$$P[Y_r = y] = \binom{y+r-1}{r-1} (1 - p)^y p^r \ .$$

The mean of this distribution is $r(1 - p)/p$ and the variance is $r(1-p)/p^2$. In Russian roulette $r = 1$ (if we consider a "success" the event that ends the player's life) and the mean is $(1 - p)/p = 1/p - 1$; since $p = 1/6$ (six chambers each equally likely, with one containing a bullet), there are generally five failures before the first success and the average time for the first success is $1/p = 6$ turns. Hence Russian roulette is not a game one is likely to play for very long. (This is contrary to the film *The Deerhunter* in which one character played Russian roulette professionally for several years.)

The negative binomial estimates the probability of a number of counts in a given number of trials until a side condition is satisfied (r events occur). The negative binomial is related to the Poisson distribution in the following way: In the Poisson distribution we count the number of events that occur at a constant rate, λ; in the negative binomial the rate parameter λ is a random variable (with a gamma distribution $r = \alpha, p = 1/ (\beta + 1)$).

To estimate p and r compute the sample mean $\bar{X}$ and variance s^2. Then method-of-moments estimates are

$$\hat{p} = \bar{X} / s^2$$

$$\hat{r} = \bar{X}^2 / (s^2 - \bar{X})$$

One problem is that this $\hat{r}$ will not be an integer.

8.4 Hypergeometric Distribution

Suppose an urn has N balls, X white and $N - X$ black. If a sample of n balls is drawn at random without replacement, then the probability of obtaining $Y = k$ white balls among the n is

$$p[Y = k] = \frac{\binom{X}{k} \binom{N - X}{n - k}}{\binom{N}{n}} \; .$$

This is a hypergeometric distribution.

If each ball were replaced after drawing, the probability of drawing a white ball would be X/N for each drawing and the distribution of Y would be binomial with parameters n and $X/N = p$. If the population can be considered infinite, the hypergeometric also reduces to the binomial. A rule of thumb is if the population succeeds 10 times the sample size, then the binomial gives a good approximation to the hypergeometric.

The mean of the hypergeometric is np and the variance is $\left[\dfrac{N-n}{N-1}\right] n \, p(1-p)$, where $p = X/N$.

8.5 Multinomial Distribution

We can generalize from the occurrence of one event, E, to a set of mutually exclusive and exhaustive events, $E_1, E_2, \ldots, E_k$. Thus, there are k categories of events. The probability of occurrence of each event E_i is p_i for each independent trial. The total number of trials is fixed at N. Let Y_i be the frequency of event E_i in N trials. Note that $\sum p_i = 1$ and $\sum Y_i = N$. These data can be arranged as a one-way contingency table:

	Categories			
	E_1	E_2	- - -	E_k
Counts	Y_1	Y_2	- - -	Y_k
Probabilities	p_1	p_2	- - -	p_k

Then the joint distribution of the Y_i's is

$$p[Y_i = y_i \, ; \, i = 1, 2, \ldots, k] = \frac{N!}{y_1! y_2! \ldots y_k!} \, p_1^{y_1} \, p_2^{y_2} \ldots p_k^{y_k}$$

8.6 Product Multinomial Distribution

Suppose we have a two-dimensional contingency table in which there are two possible sets of events $(E_1, E_2, \ldots, E_I)$ and $(F_1, F_2, \ldots, F_J)$ that can co-occur:

	F_1	F_2	- - -	F_J
E_1	m_{11}	m_{12}	- - -	m_{1J}
E_2	m_{21}	m_{22}	- - -	m_{2J}
.	.	.		.
.	.	.	- - -	.
.	.	.		.
E_I	m_{I1}	m_{I2}	- - -	m_{IJ}

Let the m_{ij} represent the cell frequencies.

If only the grand total $\sum_{i,j} m_{ij} = N$ were fixed then the joint distribution of the m_{ij} is the multinomial distribution already described:

$$p[y_{ij} = m_{ij}] = \frac{N!}{m_{11}! m_{12}! \dots m_{IJ}!} (p_{11})^{m_{11}} (p_{12})^{m_{12}} \dots (p_{IJ})^{m_{IJ}} ,$$

where p_{ij} would be estimated as m_{ij}/N.

In many studies, however, it is much more likely that there will be several groups of people and that the number of individuals sampled within each group will be fixed. This implies a fixed row total for each row (or a fixed column total for each column). This is extremely useful for observational research because the row totals will represent the frequency of a particular coding category. In most coding schemes the relative frequencies of the codes are relatively constant. For example, it is rare for married couples to summarize the partner's point of view but agreement and disagreement are relatively frequent codes. Thus, if we record a married couple for a fixed period of time, a reasonable assumption is that the marginal (row or column totals) are fixed. In this case we obtain the distribution by taking the multinomial distribution and finding the conditional distribution with fixed row or column totals.

Suppose we use the notation m_{+j} for summation within the jth column (i.e., we add rows within the jth column. For example, for a table with two rows, $m_{+1} = m_{11} + m_{21}$, sums across the two rows in the first column of the table [see Fienberg, 1978, p. 27]). If we fix the column totals, the product multinomial distribution is

$$p[y_{ij} = m_{ij} \mid y_{+j} = m_{+j}] = \prod_j \left[\frac{m_{+j}!}{\prod_i m_{ij}!} \prod_i (\frac{p_{ij}}{p_{+j}}) \right] ,$$

where $p_{ij} = m_{ij}/m_{++}$, and where $\prod$ stands for the *product* in much the same way as a $\sum$ stands for a sum:

$$\prod_{i=1}^{k} a_i = a_1 a_2 \dots a_k \; ; \sum_{i=1}^{k} a_i = a_1 + a_2 + \dots + a_k .$$

8.7 Example

Bishop et al. (1975) presented data on the number of men on base when a home run is hit for a particular year of the National Baseball League (see Table 8.1). The data are zero men on base = 421, one man on base = 227, two men on base = 96, three men on base = 21 (total men on base = 765).

One model to try is the binomial with three trials ($N = 3$). If we let p = probability a person will be on base at any one time, then no men on base is zero successes with probability $(1 - p)^3$; one man on is one success with probability $3p(1 - p)^2$; two men on is $3p^2(1 - p)$; and, three men on is p^3. Then

$$\hat{p} = \frac{1}{765} \left[\frac{1}{3}(227) + \frac{2}{3}(96) + 1(21) \right] = .21 \ .$$

The binomial thus gives the expected frequencies shown in Table 8.1.

To estimate the goodness-of-fit of this binomial model we can use the Pearson X^2,

$$X^2 = \sum \frac{(O_i - E_i)^2}{E_i} = 53.6 \ ,$$

which is significant with 3 degrees of freedom. Thus, the binomial model gives a very poor fit to the data.

The Poisson sampling model can be selected on the basis that having men on base is a rare event. The mean is

$$\hat{\lambda} = \bar{X} = \frac{0 \times 421 + 1 \times 227 + 2 \times 96 + 3 \times 21}{765} = 0.63 \ .$$

The Poisson probability model gives values far beyond three men on base, so the expected counts for $x \geq 3$ were combined. The model is reasonable if no one ever left base (got tagged out). It provided a better fit to the data (see Table 8.2).

Table 8.1

men on base	0	1	2	3
data (O_i)	421	227	96	21
binomial (E_i)	377.2	300.8	80.0	7.1

Table 8.2

men on base	0	1	2	3
data (O_i)	421	227	96	21
Poisson (E_i)	407.8	256.1	80.9	20.2

The numbers were computed as follows:

$x = 0$ $\qquad E = 765(e^{-.63})$

$x = 1$ $\qquad E = 765(.63\, e^{-.63})$

$x = 2$ $\qquad E = 765(\dfrac{(.63)^2}{2}\, e^{-.63})$

$x = 3$ $\qquad E = 765(\dfrac{(.63)^3}{6}\, e^{-.63})$

The Pearson chi square is 6.5, which, while still significant at $\alpha = .05$, provides a better fit to the data.

Although not considered by Bishop et al., the negative binomial provides an even better fit. We can compute $\overline{X} = .63$, $s^2 = .65$, $\hat{p} = .97$, $\hat{r} = 20.3$. We will truncate $\hat{r}$ to 20. This gives us

$x = 0$ $\qquad E = 765(\hat{p})^{20} = 415$

$x = 1$ $\qquad E = 765(20\, \hat{p}^{20}\, (1 - p) = 250$

$x = 2$ $\qquad E = 765(\dfrac{20 \times 21}{2}\, p^{20}\, (1 - \hat{p})^2) = 78.98$

$x = 3$ $E = 765 \, (1 - \text{sum of previous probabilities}) = 21.12.$

The Pearson chi square is 5.8, so this model gives the best fit to the data.

9

LAG SEQUENTIAL ANALYSIS

Suppose in observing the behavior of a married couple, that H_t represents a particular behavior of the husband at time t (e.g., husband negative affect), and that W_t represents a particular behavior of the wife at time t (e.g., wife negative affect). Both H_t and W_t are dichotomous, 1 if the behavior is detected and 0 if it is not. If we examine the relationship of H_t to W_t at lags k and there are n observations, we have $n - k$ pairs (W_t, H_{t+k}), $t = 1, 2, \ldots, n-k$. An extremely useful program, ELAG, written by Roger Bakeman (1983) of Georgia State University, can be used for lag sequence analysis and to compute moving time window counts for sequences. Both Sackett and Allison-Liker z-scores can be computed by it.

Binomial. Sackett (1979) suggested that we can test the relationship between H_t and W_t by comparing the conditional probability $p_{H/W} = p(H_{t+k} = 1 \mid W_t = 1)$ to the unconditional probability $p(H_{t+k} = 1) = p_H$. Recall that $\hat{p}(H_{t+k} = 1 \mid W_t = 1)$ is the proportion of times we observe the particular husband's behavior given we had observed the particular wife's behavior k units before.

Sackett suggested the statistic

$$z_B = \frac{p_{H|W} - p_H}{\sqrt{p_H(1 - p_H)/((n - k)p_W)}}$$

where $p_W = p(W_t = 1)$. He proposed that this statistic is asymptotically normally distributed with zero mean and unit variance. This is the normal approximation to the binomial distribution.

Allison and Liker (1982) criticized this statistic. They pointed out that the denominator would be correct only if p_H were the true probability and not merely the observed proportion, which is subject to sampling error. However, they were wrong. The statistic Sackett proposed is correct *asymptotically*, which was Sackett's claim, because as n increases the sample estimate of p_H approaches the population p_H. Allison and Liker's criticism is really a fine point, not at all a serious criticism of the Sackett statistic.

They proposed a statistic that is probably a better *approximation* to a standard normal distribution. Their statistic is

$$z'_B = \frac{p_{H|W} - p_H}{\sqrt{p_H(1 - p_H)(1 - p_W)/(n - k)p_W}} \, .$$

This statistic had been derived two years earlier by Gottman (1980). It is also clear that z'_B is always greater than z_B, so z_B is a more conservative statistic than z'_B. Any errors that had been made using z_B were in a conservative direction.

Hypergeometric. Wampold and Margolin (1982) argued that the hypergeometric distribution is the distribution of choice for comparing $p_{H|W}$ to p_H. They argued as follows: Suppose there are two states; call $H_t = 1$ "state i" and $W_t = 1$ "state j." Consider a sequence of behaviors containing only state i and state j. Suppose there are n_i instances of state i arranged in a line:

i   i   i   i   ...   i   (n_i instances)

There are n_i instances in the line where state j could occur after each state i:

i   j   i   j   i   j   i   ...   i   j   (n_i instances)

Suppose we want to obtain X transitions from i to j. We need to select X of the n_i spaces. These can occur in $\binom{n_i}{X}$ ways.

Similarly, if the state js are arranged in a line, there are n_j spaces occurring prior to each instance of state j that could represent a transition

i   j   i   j   i   j   i   j   ...   i   j   (n_j instances)

from i to j. There are $\begin{bmatrix} n_j \\ X \end{bmatrix}$ ways to choose X spaces to make these X transitions from i to j. There are $\begin{bmatrix} N \\ n_i \end{bmatrix} = \begin{bmatrix} N \\ n_j \end{bmatrix}$ ways to arrange the i's and the j's, and so the probability of obtaining exactly X transitions from i to j is:

$$\frac{\begin{bmatrix} n_i \\ X \end{bmatrix} \begin{bmatrix} n_j \\ X \end{bmatrix}}{\begin{bmatrix} N \\ n_i \end{bmatrix}} = \frac{\begin{bmatrix} n_i \\ n_i-X \end{bmatrix} \begin{bmatrix} n_j \\ X \end{bmatrix}}{\begin{bmatrix} N \\ n_i \end{bmatrix}} .$$

This is the hypergeometric distribution. The mean and variance of this distribution are $n_i \, n_j / N$ and

$$\frac{n_i \, n_j}{N} (1 - \frac{n_j}{N}) (1 - \frac{n_i - 1}{N - 1})$$

and a z_H statistic can be computed. In fact,

$$z_H = \frac{z_B}{\sqrt{(N - n_i)/(N - 1)}}$$

Once again, $z_H > z_B$, so z_B is a more conservative statistic.

Comparison of Assumptions. The assumption of a hypergeometric distribution for transition frequencies has some advantages and some disadvantages compared to the binomial assumption. Wampold and Margolin (1982) showed that the test can be generalized to the sum of any set of transition frequencies. For example, this is useful for assessing m_{ij} the frequency of transitions to and from i and j, $(m_{ij} + m_{ji})$, compared to an independence model. This could be used, for instance, to assess negative affect reciprocity for the *couple*, combined over spouses. In this case the mean and variance of the sum are

$2 n_i \, n_j / \, N$

and

$2 n_i \, n_j [n_i \, n_j + (N - n_i) \, (N - n_j) - N] / \, N^2 (N - 1)$

respectively. In fact, Wampold and Margolin (1982) showed that the mean and variance of any sum $(m_{ij} + m_{ik} + m_{il} + ...)$ are

$n_i (n_j + n_k + n_l + ...) / \, N$

and

$$\frac{n_i (n_j + n_k + n_l + ...)}{N^2 (N - 1)} \; (N - n_i)(N - n_j - n_k - n_l - ...)$$

respectively. Such generalizations are useful.

The assumption of a particular distribution ought to make some sense in terms of the application, although (as seen in the baseball example) this need not be the ultimate test of the use of the distribution.

The hypergeometric distribution arises from sampling without replacement from a finite population. For example, if there are N balls in an urn, m of them white and n of them black, and we select one ball at a time, the probability of a white ball on the first trial is m/N, but on the second trial it is $(m - 1)/(N - 1)$. This is how the distribution arises. If we sampled *with* replacement, the probability of a white ball would be constant on each trial and we would have a binomial distribution for x white balls.

Which distribution makes more sense for observational data? In practice the question is not useful. It would be wise to compute both z_B and z_H when these make sense. However, the binomial assumption can be compared to the hypergeometric as follows. If the event of interest is husband negative affect rather than drawing a white ball, it is hard to see why the husband having been negative reduces the total supply of negative affect available by one. It makes much more sense to assume that the proportion of negative affect is constant. Most coding systems have reasonably stable relative frequencies for their codes.

It should be noted that as long as N is reasonably large, the quantities derived from the two distributions are similar, and, in fact, close to the normal distribution. If N is large, for example, m/N and $(m - k)/(N - k)$ are usually reasonably similar.

A disadvantage Wampold and Margolin (1983) noted about the statistics they proposed is

> These statistics ... are only appropriate for formats that allow
> a succession of a behavioral state by itself.
> (p. 756)

They cannot be used for event sequence data in which a code cannot follow itself.

9.1 Deciding Between z-scores: The Type I Error Rate

Cousins et al. (1986) recently conducted a Monte Carlo study to evaluate both power and Type I error rates for four z-scores of sequential connection: the Allison-Liker z, the Yates-corrected z, a corrected z due to Overall (1980), and the arcsine z (Cohen, 1977). They varied the base rates of the given behavior and of the targeted consequent behavior, and the total number of observations (from 200 to 3,600). The results of this study are complex; replications and extensions are needed before anything definitive emerges. Nonetheless, at the .05 alpha level of significance the Allison-Liker uncorrected z

> ... provided an excellent estimate of the Type I error rate and
> was liberal in only two instances, both occurring with a sam-
> ple size of 200 behaviors. At other base rates the uncorrected
> estimate was occasionally conservative, but usually quite
> close to the nominal alpha level. (p. 13)

Conclusions changed a bit at the .01 alpha level with the z proposed by Overall giving better results, but with the "uncorrected z a close second." The more conservative Sackett z should fare even better than the Allison-Liker z. In general, these z-scores do well and power does not appear to be compromised.

9.2 Uses of Lag Sequential Analysis

When the antecedent (also called the "given" code) is studied in relation to the consequent (also called the "target") k lags later, all the codes between the antecedent and the consequent codes are ignored. If the given code is varied across all the codes in a system, a series of "z-score profiles" can be computed and graphed for each target code. These profiles can be employed to generate hypotheses about actual sequences that may be in the data.

Bakeman and Gottman (1986; Chapter 7) describe an example of using lag sequence analysis that is based on the one given in Sackett (1974). In his research with macaque monkeys Sackett used a coding system that included the codes: (1) infant active; (2) mother touch; (3) mother nurse; (4) mother groom; and (5) infant explore.

If we are particularly interested in what happens after the infant is active, the infant active code is selected as "the given" code. Next, another code is selected as "the target." For example, mother touch might be selected as the first target.

Next, we would examine the z-scores of the target at each lag from the given. Figure 9.1 suggests that mother touch is likely to occur just after infant active, but not at other lags.

We now continue selecting different target codes. Suppose the transitional probabilities for mother nurse at lag 2, mother groom at lag 3, and infant explore at lags 4 and 5 were significant ($z > 1.96$). These results are consistent with a four event sequence: infant active, mother touch, mother nurse, mother groom, infant explore. To study this possibility further we could next make mother touch the given and compute lagged z-scores for each of the other codes as targets.

It must be noted, however, that this analysis is not a complete solution. To obtain a complete solution the z-score profiles would next have to be generated with pairs of codes as givens, then triples, and so on. Lag sequence analysis is a trick to get around the problem of not having enough data for a complete Markov analysis of second or third order.

9.3 A New Lag Sequential Analysis Rule of Thumb

The problem with lag sequential analysis is that the inferences drawn about sequences are not the same as in the full model's test. We need to find a statistical test in which a particular sequence in a set of data employs only the lagged transitions and transition probabilities. For

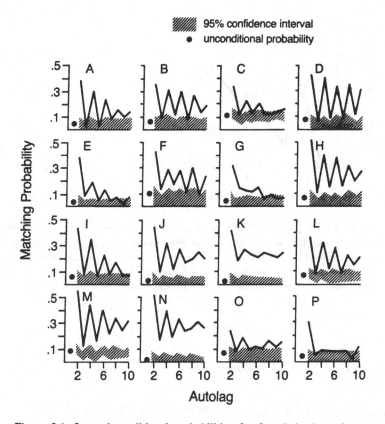

Figure 9.1. Lagged conditional probabilities for four behaviors given previous infant activity (from Sackett, 1974). Dashed lines represent simple probabilities. Because the criterion cannot follow itself, the number of behaviors that can occur at lag 1 is one less than for other lags; therefore, simple probabilities at lag 1 are somewhat higher than elsewhere.

example, if we have four codes A, B, C, and D, and an observed data set ABCAABCCDABCABDCABC, how can we test to see if ABC is a common sequence in these data using only the lagged probabilities with single codes as the givens, not pairs?

We suggest that the solution lies in computing two expected frequencies, one under the null hypothesis that the number of ABC triples can be computed by rolling forward the lag 1 model

$$E_1 = (N-2)\,p(A)\,p(B_{+1} \mid A)\,p(C_{+1} \mid B)$$

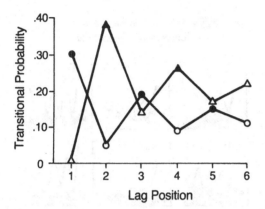

Figure 9.2. A lagged probability profile for nondistressed couples. Triangles represent transitional probabilities for Husband Complaint at the lag specified, given Husband Complaint at lag 0. Circles represent transitional probabilities for Wife Agreement at the lag specified, given Husband Complaint at lag 0. Darkened-in symbols indicate that the corresponding z-score is significant.

$$\hat{E}_1 = 17 \times .32 \times .83 \times .80 = 3.6$$

and another under the assumption that this is not the case

$$E_2 = (N - 2)\, p(B_{+1} \mid A)\, p(C_{+2} \mid A)$$

$$\hat{E}_2 = 17 \times .83 \times .67 = 9.5 \ .$$

Then compute

$$LRX^2 = 2\, E_2\, ln\, E_2 \,/\, E_1 = 2 \times 9.5\, ln\frac{9.5}{3.6} = 18.4 \ ,$$

with one degree of freedom. The test is valid to the extent that E_1 is a special case of E_2. One can sum all such chi squares for each sequence the investigator thinks is in the data. This test can be extended to any order chain.

9.4 Detecting Cyclicity

If the z-score of sequential connection rises and falls with increasing lag, it is possible that the data are cyclic and that synchronicities exist across codes. Sackett (1980) computed autolags and crosslags (see Figure 9.2) for states of sleep, awake but inactive, and active. A spectral density function based on these data would show cyclicity in them. For those who have some familiarity with time-series analysis, the z-score is related to the autocovariance function (see Gottman [1980] for an exact equation); the Fourier transform of this autocovariance function is, in fact, the spectral density function (see Gottman [1981] for a discussion of time-series analysis). This analysis is well worked out for non-discrete data. For discrete data there is some evidence that the generalization holds. See Kedem's (1980) book on binary time-series analyses.

9.5 Studying Coalitions Within Groups

Sackett (1980) studied the talk and silence patterns of 16 speakers at a conference on observational methods. He analyzed all dyads in terms of whether they facilitated or inhibited one another's talk, or had no influence. Asymmetrical and symmetrical relationships are possible in this analysis, as are cliques and coalitions. Such groupings, however, are based on interactive sequences rather than on preferences as in the usual sociograms (see Figure 9.3).

9.6 Examples

9.6.1 Ting-Toomey (1983)

Thirty-four married couples discussed their joint responses to one item on the Spanier marital adjustment scale, and one improvised conflict task described by Gottman (1979a). Couples were videotaped and the tapes were coded using Ting-Toomey's Intimate Negotiation Coding System (INCS). Three codes for each spouse: integrative (e.g., agree); descriptive (e.g., describe problem); disintegrative (e.g., disagree). *Markov chain analyses* were performed.

1. **Homogeneity.** The two discussions were compared for a random 10% of the couples. The LRX^2 was nonsignificant.

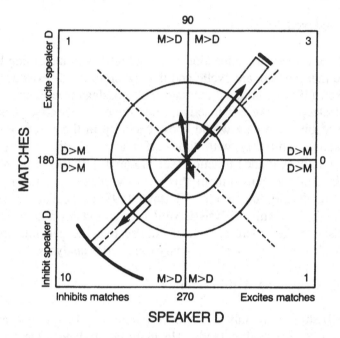

Figure 9.3. Polar graph of Speaker D average z = sum vectors and angles with all other people. In the right half D as criterion *is followed by* other people at probabilities above expected values (excitatory dependency), while in the left half D *is followed by* others less than expected (inhibitory dependency). In the upper half Speaker D *follows* other people more than expected, while in the lower half D *follows* others less than expected. The number of people involved in each relationship type are shown in the quadrant corners. Arcs near the corner measure the range of vector angles over the N individuals in each quadrant (no arc with $N = 1$). Vector lengths measure average strength of relationship in z = sum units (inner circle = 2, outer circle = 4). Rectangles give the standard deviation of vector length. In areas 45 degrees above and below the 0-180 plane, the Speaker D contribution to the relationship is stronger than that of other people. In areas 45 degrees to the right and left of the 90-270 plane the Speaker D contribution is weaker than that of other people. (From Sackett, 1980)

2. **Stationarity and Couple Homogeneity.** The stationarity LRX^2's for each group were not significant. The homogeneity analyses for each group were also not significant.

3. **Order.** Tested zero versus first order and got significant LRX^2's for each of the three adjustment groups. It uses a first-order process. The high and medium marital adjustment groups were compared, $LRX^2 = 260.0$, $df = 210$, $p < .05$; the low marital and medium

$(LRX^2 = 278.7, df = 210, p<.01)$ and the high and low marital adjustment groups $(LRX^2 = 522.0, df = 210, p<.01)$.

Lag sequential analyses were employed to study specific patterns (up to lag 5). Some conclusions about the high marital adjustment (HMA) group are: (1) The act of coaxing in lags 2, 4, and 5 was found to be sequentially dependent on the criterion code in the HMA group, whereas no such pattern was found in the LMA group; (2) the act of confirming in lag 2 was found to be statistically significant on the act of confirming in the HMA, while no such pattern was found in the LMA group; (3) the act of socioemotional description was found to be dependent on the criterion of confirming in the HMA group. Ting-Toomey wrote:

> Overall, the results of the HMA interaction indicated that the negotiation processes of these high adjusted couples were represented by the stripes of verbal coaxing, confirming, socioemotional and task oriented strategies ... It seems that a low-adjustment marital interaction, when one spouse starts to verbally attack the other spouse directly (confront) or indirectly (complaint), the other most often responds with defensive messages. While the low-adjusted couples did engage in such acts as socioemotional questioning or task-oriented questioning, no significant sequential pattern was found as stemming from the INCS integrative and descriptive categories. All significant sequential patterns were derived from the INCS disintegrative categories. (p. 314)

9.6.2 Bakeman and Adamson (1983)

Infants 6 to 18 months old were observed (longitudinally) in their homes, playing with their mothers and with peers. Sequential analyses employed z-scores within subjects and analyses of variance were conducted on the z-scores to study contextual effects. All developmental trends observed were similar regardless of partner: Person engagement declined with age while coordinated joint engagement increased. Sequential analyses showed that both passive and coordinated joint engagement were likely to be preceded by object play, which was a likely consequent of coordinated joint engagement. The authors summarize:

In summary, communicating with others about objects demands attention to both social and object aspects of one's surroundings . Moreover, it is a process that may well depend on the privacy of social relationships, which builds upon skills nurtured during the earlier period of face-to-face play, and which expands during the emergence of a "triadic" (infant-object-mother) interactive system. (p. 13, ms.)

9.6.3 Margolin and Wampold (1981)

Twenty-two distressed and 17 nondistressed married couples were video-taped discussing two currently conflictual topics. Data were coded with the Marital Interaction Coding System (MICS) and further summarized into a positive/neutral/negative trichotomy for each spouse.

Nonsequential results. Margolin and Wampold found significant differences between groups in unconditional probabilities (base rates). Distressed couples were more likely than nondistressed couples to engage in problem solving, verbal positive, and nonverbal positive codes. Within these categories univariate analyses yielded significant differences on problem solution, agree, assent, physical positive, and smile/laugh. There was more neutral behavior in nondistressed couples (command, problem description, and interrupt).

Wives were more affective than husbands, both more positive (only for smile/laugh) and more negative (complain and criticize). No interactions between sex and distress were significant.

Sequential analyses. Margolin and Wampold performed couple-by-couple analyses on the z-scores used to index three sequences (see Table 9.1): positive reciprocity, negative reciprocity, and negative reactivity (defined as the likelihood that a positive consequent following a negative antecedent will be *less* than the unconditional probability of positive behavior).

Conclusions were that positive reciprocity characterizes *both* groups of couples, with no significant differences between distressed and nondistressed couples. Both for negative reciprocity and negative reactivity the two groups of couples were significantly different, as predicted.

Margolin and Wampold also found that positive reciprocity was correlated with positive base rates ($r = .53$, $p < .01$) but that negative reciprocity and negative behavior were uncorrelated ($r = .02$). Negative

Table 9.1. Comparison of Couple-by-Couple z-scores by Margolin
and Wampold (1981)

Group	Mean		t	t-one sample
		Lag 1 z-scores		
		Positive reciprocity		
Distressed	1.07		1.17	2.93**
Nondistressed	1.70			4.35*
		Negative reciprocity		
Distressed	0.75		1.91*	1.87**
Nondistressed	−0.12			−.57
		Negative reactivity		
Distressed	−.52		1.95*	−2.59**
Nondistressed	.12			.44

$* P < .05$ $** p < .01$

Table 9.2. Conditional Probabilities of Empathic and Aversive
Responses to Complaints

Marital Adjustment	Vague Complaint→ Legitimize/Empathize	Focused Complaint→ Legitimize/Empathize
Moderate	.16**	.09
High	.18*	.13
Very High	.10**	.19**

	Vague Complaint→ Negative Behavior	Focused Complaint→ Negative Behavior
Moderate	.11	.15
High	.14	.08
Very High	.05	.02

Conditional probabilities were tested against unconditional using the AL z-score
$(* p < .01 ;   ** p < .001)$

reactivity was correlated with negative behavior ($r = .41$, $p < .01$), but not positive behavior ($r = .10$, n.s.).

> Interestingly, the negative reciprocity index shows minimal overlap with negative base rates. In other words, negative reciprocity is not related to the actual frequency of negative behaviors but, as seen in earlier results, occurs as a function of whether couples perceive themselves to be maritally distressed. (p. 563)

It is interesting that the reciprocity of negative affect was independent of the amount of negative affect. Sequential analysis provides new information about marital interaction over and above what would have been obtained by collapsing over time.

9.6.4 Cousins and Vincent (1983)

This was a study of couples undergoing a major life transition, the birth of the first child. Forty-two couples were videotaped discussing a situation *outside the marriage* in which they had been emotionally upset. Data were collected one month after birth of the child. Most couples were happily married, so this was a comparison of those "moderately," "highly," or "very highly" satisfied with their marriages. The five minutes of conversation were coded with the Marital Interaction Coding System (MICS).

We will summarize their results for the patterns of contingency following complaints (see Table 9.2). All couples showed a significant increase in legitimize/empathize behavior following vague complaint ($z = 4.80, 2.98, 3.37$). Only couples with the highest adjustment scores followed a focused complaint significantly more often than base rate with legitimize/empathize ($z = 6.22$).

To test for group differences the highest and lowest adjustment groups were compared using the z-score estimate of the Mann-Whitney U, corrected for continuity and ties. Couples with the lowest adjustment were more likely to follow complaints with negative behavior than couples high in adjustment ($z = 2.21, p < 0.05$). They wrote:

> Skillful partners seemed to be better able to detach themselves from negative affect expressed by their spouse and refrain from critical, sarcastic, or defensive comments than less

skillful partners. There was modest support for the value of positive behavior. (p. 681)

9.6.5 Revenstorf et al. (1981)

Revenstorf et al. (1981) coded videotapes of conflict discussions of ten couples in marital therapy (T) and 10 normal (N) couples in a study conducted in Munich. Data were coded with the Marital Interaction Coding System (MICS). Marginal distributions showed that T couples were more negative and less positive than N couples. An analysis of sequences showed that interaction patterns were also different in the two groups of couples. They wrote:

> ... of particular interest here is how the spouses react to a problem description (P) of the partner, because this is where quarrels begin. Comparing these conditional distributions for both samples yields statistical significance. T-couples react more negatively and less positively then N-couples ($X^2 = 29$, $df = 1$). (p. 5, ms.)

Sequences that indexed attempts at reconciliation and acceptance were far more frequent in the normal sample. Devaluations, fighting back, and yes-butting were more frequent in the therapy sample. After a negative statement no immediately predictable sequence was obtained for normal couples, whereas in therapy couples both partners continued to reciprocate negatively. See Figure 9.4 for a summary of group differences in long chains of negative affect.

9.7 Independence Assumption in Computing z-Scores

Bakeman and Dorval (1988) raised the issue of the nonindependence of codes in the computation of frequencies when the moving time window is used. For example, if there are five codes, A, B, C, D, and E, and the observed sequence is ABDCE, then a moving time window computes pairs as follows: (AB)DCE; A(BD)CE; AB(DC)E; and ABD(CE). Clearly the counts are not independent, since the last element of a pair becomes the first element of the next pair. The problem is eliminated if only non-overlapping pairs are employed. There are two non-overlapping sets of pairs: 1-2, 3-4, ...; and 2-3, 4-5, ... It is possible to form a table for both the overlapping and the two non-overlapping *r*

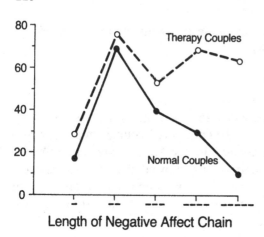

Figure 9.4. Probabilities of long chains of negative affect (from Revenstorf et al. 1981).

computations. Bakeman and Dorval conducted a Monte Carlo study for equiprobable coding categories and one in which the unconditional probabilities of the codes varied. There were 1,000 codes in each of the series they generated, which provided 500 tallies for the non-overlapped computation and 1,000 for the overlapped computation. For each type of series 100,000 trials were performed and the distributions of the computed z-scores were compared for the two methods using a Kolmogorov-Smirnov test, adjusted for multiple comparisons. The distributions were not significantly different. They concluded:

> The Monte Carlo simulations indicate that, although the assumption of sampling independence seems violated when overlapping pairs are tallied, this violation is in fact inconsequential. (p. 10, ms.)

Part III

The Timetable and the Contextual Design

10

LOG-LINEAR MODELS

It is not enough to be able to say that the order of sequential dependency is two or three, or even that there is some temporal structure in the data. In most of our research we have an experimental design and want to know if *specific* sequential patterns vary with our experimental factors. For example, suppose we have studied younger and older children and observed interactions of pairs of children either with their best friends or with strangers. We may want to know if friends are more likely than strangers to clarify their message in response to a clarification request. This is the kind of question sequential analysis is designed for.

Log-linear analysis is the generalization of analysis of variance to contingency tables in which the entries in each cell are counts, sampled under some distributional assumption. The typical case of interest to us is one in which we have some experimental design (which we call *the contextual design*) and in each cell of this design we have a timetable. The timetable has been constructed to be of a particular order based on our order and stationarity tests. It has also been based on pooling data across subjects (dyads, families, groups, etc.) based on our homogeneity tests. We now want to know how particular cells in the timetable that are of theoretical interest vary as a function of main effects and interaction effects in the contextual design. Such questions can be answered with *log-linear models*.

In Chapter 13 we discuss the use of *logit models*, which are the generalization of regression techniques. Logit analysis is particularly useful when we conceptualize our data in terms of one dichotomous response variable.

Table 10.1 illustrates a typical situation. We have two codes, *A* and *B* and the timetable is repeated for each cell in our contextual design. The data are arranged to resemble a contingency table. In our case the data consist of counts or frequencies, using the moving time window.

Table 10.1. Contextual design with timetable

Young best friends

$t + 1$
A   B

A
t
B

Young strangers

$t + 1$
A   B

A
t
B

Older best friends

$t + 1$
A   B

A
t
B

Older strangers

$t + 1$
A   B

A
t
B

Table 10.2. Notation for log-linear models

			Factor 2 (Consequent)		
			LEVELS		
			1	2	Marginal frequencies
Factor 1	L E V E L S	1	x_{11}	x_{12}	x_{1+}
(Antecedent)		2	x_{21}	x_{22}	X_{2+}
	Marginal frequencies		x_{+1}	x_{+2}	N

Usually, of course, there will be more than two codes, such as codes $A_1, A_2, \ldots, A_S$. If there are two children in each of our interaction sessions, we will obtain a set of codes for each child. For example, if we record only talk (T) or silence (S) for each child (1 or 2), we have four codes, $A_1 = T1$, $A_2 = S1$, $A_3 = T2$, $A_4 = S2$, for talk or silence, respectively, for child 1 or child 2.

The timetable repeated in each cell of the design can, of course, be more than two-dimensional, depending on the order of the Markov process, as determined by the methods of Part 1.

How do we analyze our data to answer the general question of how the factors in the experimental design affect the frequency of our codes and their sequential structure? The answer is that we will employ a method called *hierarchical log-linear models*. The notation in log-linear models will recall the analysis of variance.

10.1 Notation for Log-Linear Models

We need to introduce some new notation (see Table 10.2). "Factor 1" is the antecedent. It has two "levels," the two codes. "Factor 2" is the consequent. It also has two "levels," the two codes. The symbol m_{i+} denotes the frequency with which the antecedent code was i, and m_{+j}, that the consequent code was j. The plus indicates that we have summed over this factor. We will define m_{ij} = expected frequency and x_{ij} = observed frequency. The independence model can be written in terms of the estimated expected value in the (i, j) cell:

$$\hat{m}_{ij} = x_{i+}\, x_{+j}/N \tag{10.1}$$

Once again, independence means that there is no relationship between antecedent and consequent codes. If we take natural logarithms of both sides of Equation (10.1), the product becomes a sum

$$\log \hat{m}_{ij} = -\log N + \log x_{i+} + \log x_{+j}$$
$$= \hat{u} + (\hat{u}_1)_i + (\hat{u}_2)_j \tag{10.2}$$

A bit of explanation is necessary to explain the notation employed in the last line of Equation (10.2). The symbol $(u_1)_i$ represents the mean corresponding to the first (antecedent) factor at level (i.e., code) i. The symbol $(u_2)_j$ represents the mean corresponding to the second (consequent) factor at level (i.e., code) j. In general, there will be I levels of factor 1 and J levels of factor 2.

The similarity of Equation (10.2) to analysis of variance notation makes it possible to write the logarithm of the parameter $\hat{m}_{ij}$ as

$$\log \hat{m}_{ij} = \hat{u} + (\hat{u}_1)_i + (\hat{u}_2)_j$$

in which

$$\hat{u} = \frac{1}{IJ} \sum_i \sum_j \log \hat{m}_{ij}, \text{ the grand mean}$$

$$\hat{u} + (\hat{u}_1)_i = \frac{1}{J} \sum_j \log \hat{m}_{ij}$$

$$\hat{u} + (\hat{u}_2)_j = \frac{1}{I} \sum_i \log \hat{m}_{ij}$$

In this notation $(\hat{u}_1)_i$ and $(\hat{u}_2)_j$ represent deviations from the grand mean so that we have analogous side conditions so familiar from the analysis of variance: $\sum_i (u_1)_i = \sum_j (u_2)_j = 0$.

Equation (10.2) in English says that the effects are additive row and column effects and not interactive. This is an extremely boring model for us because it states that there is no sequential structure at all.

How would we write a model that *did* have sequential structure? We need an ij term. Such a model is given by Equation (10.3),

$$\hat{m}_{ij} = x_{i+} \, x_{+j} \, s_{ij} \tag{10.3}$$

in which s_{ij} is a "correction factor" for nonindependence. This general model is also called the *saturated model*. If we take logs again:

$$\log \hat{m}_{ij} = \hat{u} + (\hat{u}_1)_i + (\hat{u}_2)_j + (\hat{u}_{12})_{ij} \tag{10.4}$$

This last term, $(\hat{u}_{12})_{ij}$, can be interpreted to read that nonindependence produces a two-way interaction term. We will be interested in testing whether $(\hat{u}_{12})_{ij} = 0$ for all i and j.

Table 10.3. Three-way table

					Factor 3						
		$(3)_1$				$(3)_2$				$(3)_3$	
	$(2)_1$	$(2)_2$	$(2)_3$		$(2)_1$	$(2)_2$	$(2)_3$		$(2)_1$	$(2)_2$	$(2)_3$
$(1)_1$				$(1)_1$				$(1)_1$			
$(1)_2$				$(1)_2$				$(1)_2$			
$(1)_3$				$(1)_3$				$(1)_3$			

So far, we've been pretty cagey; we haven't said *how* we will perform these tests, but readers will have guessed that likelihood ratio chi-squared statistics enter into it in some way. First, we will continue talking in terms of the notation, just so we can become familiar with it.

10.2 Three-way Classification

If we go back into time more than one time unit to study the nature of time dependence among codes, we create a *three*-dimensional contingency table. In this case we count trigrams. For example, for the observed string 112332221 the first trigram is 112, the second is 123, the third is 233, the fourth is 332, the fifth is 322, the sixth is 222, and the seventh is 221. A good method for writing a three-way table is shown in Table 10.3. Note that the entries in this table are not independent counts.

For this table we can write the independence model as follows

$$\hat{m}_{ijk} = x_{i++} \, x_{+j+} \, x_{++k}/N^2 \tag{10.5}$$

where x_{++k} now represents the third factor. Once again, taking logarithms and employing our notation, we obtain:

$$\log \hat{m}_{ijk} = \hat{u} + (\hat{u}_1)_i + (\hat{u}_2)_j + (\hat{u}_3)_k \tag{10.6}$$

for the independence model. We will now consider special models.

10.3 Playing with the Notation

We shall postpone the discussion of statistical procedures for a while and talk only about models so that readers can become comfortable with the new notation. Readers should pretend it is easy to test the significance of each term in our models to avoid anxiety about the delay in discussing statistical testing. We shall always refer back to the three-way table (Table 10.3).

10.3.1 Case (1)

i is independent of j and k. This means that *j* and *k* may be related, but *i* is not related to either of them. This could imply that the time dependence in the data extends to dyads but not to triads. The word "could" is in the last sentence only because the time dependence might be complicated; for example, it might skip the first time point but include a later one. This would be represented

$$\hat{m}_{ijk} = \frac{1}{N^2} \, x_{i++} \, x_{+j+} \, x_{++k} \, t_{jk} \tag{10.7}$$

where t_{jk} is a "fudge factor" that intentionally includes only *j* and *k* and not *i*. Taking logs again:

$$\log \hat{m}_{ijk} = \hat{u} + (\hat{u}_1)_i + (\hat{u}_2)_j + (\hat{u}_3)_k + (\hat{u}_{23})_{jk} \tag{10.8}$$

In analysis of variance language this model reads that there is interaction only between factors 2 and 3.

10.3.2 Case (2)

i is independent of j, conditional on k. This means that once *k* is specified, for each level of *k*, *i* and *j* are independent. However, the form of the independence can be different at different levels of *k*. A numerical example will help explain this concept. Table 10.4 was constructed so that for each level of *k*, *i* and *j* are independent. For example, for $k = 1$, the first cell entry $2 = (3 x 6)/9$. This model can be written:

Table 10.4. Example of a special three-way model
(i is independent of j, conditional on k).

	$k = 1$				$k = 2$		
	$(2)_1$	$(2)_2$			$(2)_1$	$(2)_2$	
$(1)_1$	2	1	3	$(1)_1$	12	4	16
$(1)_2$	4	2	6	$(1)_2$	6	2	8
	6	3	9		18	6	24
	2:1				3:1		

$$\hat{m}_{ijk} = \frac{1}{N^2(k)} \, x_{i+k} \, x_{+jk} \qquad (10.9)$$

Equation 10.9 states that N varies with k. In Table 10.4, N_1 is 9 and N_2 is 24. It also states that for each fixed level of k the rows (i) and columns (j) are independent. This model can be rewritten using two "fudge factors" as follows:

$$\hat{m}_{ijk} = \frac{1}{N^2} \, x_{i++} \, x_{+j+} \, x_{++k} \, t_{ik} \, u_{jk} \qquad (10.10)$$

The fudge factor t_{ik} represents the fact that i depends on k, and the fudge factor u_{jk} represents the fact that j depends on k. Note the absence of a fudge factor that represents the dependence of i on j. Taking logs:

$$\log m_{ijk} = \hat{u} + \hat{u}_1 + \hat{u}_2 + \hat{u}_3 + \hat{u}_{13} + \hat{u}_{23} \qquad (10.11)$$

Two terms are out of this model, a 12 interaction term (u_{12}) and a three-way 123 interaction term (u_{123}). Note that the question of the existence of a three-way interaction term is salient to Markov modeling. It is "Are times 1 and 3 still dependent if time 2 is partialed out?"

Warning. Table 10.4 also illustrates that it is possible for i and j to be independent at every level of k, and yet for the table of i and j, collapsed over the levels of k to obscure this fact.

Table 10.5. The dangers of collapsing over a variable
(from Upton, 1978)

	C_1				C_2		
	B_1	B_2	Total		B_1	B_2	Total
A_1	15	5	20		28	12	40
A_2	15	5	20		42	18	60
Total	30	10	40		70	30	100

	B_1	B_2	Total
A_1	43	17	60
A_2	57	40	80
Total	100	40	140

Table 10.6. Illustration of Simpson's paradox
(from Upton, 1978)

	C_1				C_2		
	B_1	B_2	Total		B_1	B_2	Total
A_1	95	800	895		400	5	405
A_2	5	100	105		400	195	595
Total	100	900	1000		800	200	1000

	B_1	B_2	Total
A_1	495	805	1300
A_2	405	295	700
Total	900	1100	2000

Upton (1978) discussed the example in Table 10.5. In this top panel of Table 10.5, A and B are independent at every level of C, but if only the bottom panel were examined, we would conclude that A and B are not independent. The BC table would show that B and C were not independent, and the AC table would show that A and C were not independent. This example points out the disadvantages of analyzing data from a three-way table using only all two variable tables. An even greater anomaly is known as *Simpson's paradox* (discovered by Simpson, 1951). First, a definition is necessary. Positive association (in 2×2 tables) between variables A and B is said to exist if the quantity $X_{11} X_{22} - X_{12} X_{21}$ is < 0 and a negative association is said to exist if the quantity is ≥ 0. In Table 10.6, A and B display positive association for each of the two levels of C, and negative association See Upton (1978, p. 43) for more discussion.

10.4 Hierarchical, Non-Hierarchical and Conditional Nested Models

By far the most commonly used type of log-linear model is the so-called hierarchical model, one that involves main effects and their multiplicative interactions – *in hierarchical models, all of the main effects constituting any interaction term must be included in the model before the interaction term itself may be included, as well as all lower-order interactions (of a higher-order interaction).* In non-hierarchical models, on the other hand, interaction terms may be included even if one or more of the constituent main effect terms have been omitted.

Another wholly new class of models – "conditional (or nested) models" – have lately come into vogue in the contingency tables analysis literature (cf., Magidson, Swan, & Berk, 1981). Such models provide a specific approach to transforming multiplicative interactive terms into substantively more meaningful multiple main effect terms. They create conditional main effects for one variable that are nested within each level of another explanatory variable.

Conditional models account for the exact variance as hierarchical models do, if they include equivalent terms. However, conditional or nested models have one major advantage: They allow multiplicative interaction terms to be eliminated, thereby simplifying the job of meaningful interpretation. Moreover, nonsignificant terms may be dropped and the model reduced thereby to lead further to substantive insights.

The major disadvantage with conditional models is that the researcher has to assume that one independent variable is logically or theoretically superior to the variable(s) nested within it. This assumption is analogous to the assumption of unidirectional causation that is made with recursive models in structural equations modeling. Sometimes, it may indeed be possible to have theoretical insights and/or empirical evidence regarding whether one variable is logically superior to another.

There is no theoretical reason as to which model one uses with the various estimating algorithms available. In practice, however, one is limited by computer program(s) available and what they can do.

10.5 Statistical Tests

We will compute a special *series* of statistics, each called G^2, the *likelihood ratio statistic* (also called LRX^2). Each G^2 is distributed asymptotically as chi square, and is computed as

$$G^2 = 2 \sum_i (observed)_i \log \frac{(observed)_i}{(expected)_i} \; .$$

Note that we cannot simply compute G^2 from a table of observed counts. Why not? Because we need some way of getting the expected counts, and to do that *we need a model, which is a procedure for obtaining expected counts.*

Usually we begin with our most complex table, the data, and fit the "fully saturated model," which has all the main effect and interaction terms in it. We have $\hat{m}_{ijk} = x_{ijk}$ in a fully saturated model. We then compare this observed set of counts to a hierarchical series of models that essentially drop terms out of the model. In our discussion of log-linear models, three types were reviewed. For three factors (i.e., a three-dimensional table), *dependence* existed between Factors 1 and 2 if there was a u_{12} term in the model. This meant that Factors 1 and 2 are directly linked. Factors 1 and 2 being independent conditional on Factor 3 (*conditional independence*) was a second type of model. In this case u_{13} and u_{23} were in the model, u_{12} and u_{123} were not. This is an indirect linkage. The third type of model was the *independent model*. In a first-order Markov model, for example, there would be only first-lag dependence, so the only terms in the model are u_{12}, and u_{23} and the grand mean, u, and the main effect terms u_1, u_2, and u_3.

10.6 Getting Expected Counts

The most common method for finding expected counts is *iterative proportional fitting* (IPF). Computer programs exist (using either the Deming & Stephan IPF algorithm [1940] or the generalized iterative scaling procedure of Darroch & Ratcliff [1972]) for this computation (e.g., Fay & Goodman's [1972] ECTA), but first the model must be specified. In general, the way IPF works is that if there is a u_J term in the model, the J margin is examined. For example, if there is a u_1 term, we would examine the 1 margin (summing across 2 and 3) and the expected data would have to agree with this margin. Consider an example in which the model has a 12, a 13, and a 23 term, and assume that the data are as in Table 10.7, top panel.

The iteration starts with all expected cell counts set equal to 1 (see Table 10.8, top panel). Next the AB margin is fit by determining the correction factor that each first iteration number in the AB margin must be multiplied by to make it agree with the actual AB margin. These correction factors are then multiplied by the first iteration numbers (all 1) to obtain the second iteration (second panel of Table 10.8). Next the BC margin is fit, correction factors are again computed, and a third iteration is calculated. For example, the 4/6 correction factor in the third panel of Table 10.8, cell (C_1, B_1), will multiply all the $(C_1 B_1)$ estimates in the second iteration regardless of the level of A. If the reader works through this problem numerically, the procedure will become clearer. The process is repeated until convergence is obtained.

What is this procedure accomplishing? It is finding expected counts with the triple interaction term out of the model. Hence, the saturated model (Table 10.7) is compared to the expected counts under the model of no three-way interaction. If we assume that convergence has already taken place by the bottom of Table 10.8, we would compute G^2 as in Equation (10.12), with the expected values supplied by the iterative proportional fitting algorithm.

It is important to note that there is not just one G^2, but a different G^2 for each model. Here is where the statistics come in. Suppose we have two models, A and B. In the special case that B is a smaller model than A (i.e., it is A with some terms missing), it is also the case that the difference in the two G^2's is asymptotically distributed as chi square, with degrees of freedom equal to the difference in the degrees of freedom of the A and B models. This fact makes it useful to fit a hierarchical series of models to the data.

Table 10.7. Example for iterative proportional fitting

	$C = 1$				$C = 2$		
	B_1	B_2			B_1	B_2	
A_1	1	2	3	A_1	7	9	16
A_2	3	5	8	A_2	1	2	3
	4	7			8	11	

AB margin: Add over C (this is for the 12 term)

	B_1	B_2
A_1	8	11
A_2	4	7

BC margin: Add over A (this is for the 23 term)

	B_1	B_2
C_1	4	7
C_2	8	11

AC margin: Add over B (this is for the 13 term)

	A_1	A_2
C_1	3	8
C_2	16	3

Table 10.8. Iterations for iterative proportional fitting

Start iteration

	$C = 1$				$C = 2$	
	B_1	B_2			B_1	B_2
A_1	1	1		A_1	1	1
A_2	1	1		A_2	1	1

First fit AB margin (that is, start by adding over C.)

	B_1	B_2			B_1	B_2	
A_1	2	2	Multiply each term by the appropriate correction factor → (see Table 10.7)	A_1	8/2	11/2	This gives Second iteration →
A_2	2	2		A_2	4/2	7/2	

	$C = 1$				$C = 2$	
	B_1	B_2			B_1	B_2
A_1	8/2	11/2		A_1	8/2	11/2
A_2	4/2	7/2		A_2	4/2	7/2

Next fit the BC margin (that is, sum over A)

	B1	B2			B1	B2	
C1	12/2=6	18/2=9	To make this fit with observed BC marginals must → multiply accordingly (see Table 10.7)	C1	4/6	7/9	Third iteration
C2	12/2=6	18/2=9		C2	8/6	11/9	

	$C = 1$				$C = 2$	
	B1	B2			B1	B2
A1	(8/2)(4/6)	(11/2)(7/9)		A1	(8/2)(8/6)	(11/2)(11/9)
A2	(4/2)(4/6)	(7/2)(7/9)		A2	(4/2)(8/6)	(7/2)(11/9)

	$C = 1$				$C = 2$	
	B1	B2			B1	B2
A1	2.67	4.28		A1	5.33	6.72
A2	1.33	2.72		A2	2.67	4.28

Table 10.8. Iterations for iterative proportional fitting (continued)

Next fit the AC margin

	C1	C2
A1	6.5	12.05
A2	4.05	6.95

To make this fit with observed marginals must multiply accordingly

	C1	C2
A1	3/6.95	16/12.05
A2	8/4.05	3/6.95

C = 1

	B1	B2
A1	2.67(3/6.95)	4.28(3/6.95)
A2	1.33(8/4.05)	2.72(8/4.05)

C = 2

	B1	B2
A2	5.33(16/12.05)	6.72(16/12.05)
A2	2.67(3/6.95)	4.18(3/6.95)

which is

C = 1

	B1	B2
A1	1.15	1.85
A2	2.63	5.37

C = 2

	B1	B2
A1	7.08	8.92
A2	1.15	1.85

10.7 Model Building and Testing

Beyond examining each cell, we can compare models directly once we derive the expected counts. One way to proceed is to begin with the saturated model and take out higher-order interaction terms, one at a time and determine which combinations are rejected as not significant.

Hierarchical Model Testing. Suppose we have three models, say A, B, and C, such that model A has the fewest parameters, model B has all the parameters in A, plus a few extra, and model C has all as in B, plus a few more. Then we can compute the likelihood ratio X^2 goodness-of-fit statistics, $G^2(A)$, $G^2(B)$, and $G^2(C)$; the difference $G^2(B) - G^2(A)$ (also called "conditioning on B") is also a chi-squared statistic. *Moreover, this difference has a X^2 distribution with degrees of freedom associated with the number of added parameters.* A similar set of statements holds for $G^2(C) - G^2(B)$. Thus, we can measure the *improvement of fit* as well as the goodness of fit of each model.

We will now see how to compute the degrees of freedom for a model.

Table 10.9. Terms in the fully saturated log-linear model

Term	# Independent terms	Computational formula
u	1	1
$(u_1)_i$	4	$I - 1$
$(u_2)_j$	4	$J - 1$
$(u_3)_k$	4	$K - 1$
$(u_{12})_{ij}$	16	$(I - 1)(J - 1)$
$(u_{13})_{ik}$	16	$(I - 1)(K - 1)$
$(u_{23})_{jk}$	16	$(J - 1)(K - 1)$
$(u_{123})_{ijk}$	64	$(I - 1)(J - 1)(K - 1)$
Total	125	

10.8 Computing Degrees of Freedom

Suppose there is a three-dimensional table with 5 levels of each factor. Table 10.9 presents the fully saturated model. There are 125 terms in the saturated model. The independence model has only u, u_1, u_2, and u_3, or 13 terms. This gives $125 - 13 = 112$ degrees of freedom when compared to the fully saturated model. The model that 1 and 3 are independent, conditional on 2, has the terms u, u_1, u_2, u_{12}, u_{23}, which has 45 terms, so that a comparison with the saturated model involves 80 degrees of freedom.

If some cells are zero by the logic of the experimental design (e.g., code 2 can *never* follow code 1 by definition, or a code can never follow itself), this will yield an incomplete table. In this case the iterative fitting technique starts with zeros instead of 1's in these cells, and 1 degree of freedom must be subtracted for every cell forced to zero in the model.

The number of parameters that cannot be estimated is usually the number of zeros on the fitted marginal tables (but see the discussion on pp. 115-116 of Bishop et al. [1975]). The adjustment discussed in Bishop et al. is appropriate if there are not too many empty cells, but the method is not appropriate in *very* sparse tables (e.g., if half the cells are zero, do not believe this adjustment).

10.9 Model Fitting Step by Step

It is very important to point out that the goal in model fitting is to find a simple and interesting model that fits the data, in other words, that produces a *nonsignificant* chi square, which is not significant because the model provides a good approximation to the table.

Perhaps the most commonly used technique for finding a model is to evaluate a series of models arranged in a hierarchy so that each model is contained in a previous model. This is not the only set of models possible, but the hierarchical approach has some definite advantages.

It has become standard notation to write a model using brackets, as follows: If we have three factors, A, B, and C, a model that includes only A and B and their interaction would be written [AB] and the model would be

$$\log \hat{m}_{ijk} = \hat{u} + (\hat{u}_A)_i + (\hat{u}_B)_j + (\hat{u}_{AB})_{ij} \qquad (10.13)$$

The m_{ijk} are the cell frequencies and the u's estimate the effects of each factor. Note that the presence of the interaction term implies the presence of the main effect terms. This is what makes the models hierarchical. In other words, the model in Equation (10.13) contains the following simple main effect model:

$$\log \hat{m}_{ijk} = \hat{u} + (\hat{u}_A)_i + (\hat{u}_B)_j \qquad (10.14)$$

Table 10.10 illustrates the 19 possible models for a three-way experimental design. Here is the beauty of the hierarchical arrangement: *We can evaluate the effect of a term in the model by subtracting the G^2 for the models.* For example, if the LRX^2 model represented by Equation (10.13) – remember, this uses IPF to compare (10.13) with the saturated model, that is, the observed values – is 44.20 with 24 df and the G^2 for model (10.14) is 200.70 with 39 df, then the G^2 for the AB interaction term is $(200.70 - 44.20) = 156.50$ with $(39 - 24) = 15$ df. This shows that the $(\hat{u}_{AB})_{ij}$ term must probably be included in the model.

In this manner each term in the full model can be evaluated. To summarize, each model yields its own likelihood ratio G^2 term and df. The difference between the G^2 and the df's for two suitably selected models can produce an assessment of the significance of a term in the full model. For example, in a two-dimensional table if the [A] [B] model

Table 10.10. Nineteen possible models for a three-way table

Defining set of parameters	Parameters in the model							
	u	A	B	C	AB	AC	BC	ABC
[ABC]	√	√	√	√	√	√	√	√
[AB] [AC] [BC]	√	√	√	√	√	√	√	
[AB] [AC]	√	√	√	√	√	√		
[AC] [BC]	√	√	√	√		√	√	
[BC] [AB}]	√	√	√	√	√		√	
[A] [BC]	√	√	√	√			√	
[B] [AC]	√	√	√	√		√		
[C] [AB]	√	√	√	√	√			
[BC]	√		√	√			√	
[AB]	√	√	√		√			
[AC]	√	√		√		√		
[A] [B] [C]	√	√	√	√				
[A] [B]	√	√	√					
[A] [C]	√	√		√				
[B] [C]	√		√	√				
[A]	√	√						
[B]	√		√					
[C]	√			√				
[u]	√							

$$\log \hat{m}_{ij} = \hat{u} + (\hat{u}_A)_i + (\hat{u}_{B_j}) \tag{10.15}$$

yields $G^2 = 6.9$ with $1 = $ df and the submodel [A]

$$\log \hat{m}_{ij} = \hat{u} + (\hat{u}_A)_i \tag{10.16}$$

yields $G^2 = 41.7$ with $2 = $ df, then the difference in the G^2's evaluates the $\hat{u}_B$ term as $41.7 - 6.9 = 34.8$ with df $= 2 - 1 = 1$. This would suggest that we need to keep the $\hat{u}_B$ term in the model.

10.10 Selecting a Model by Screening

Now here's the rub: In contingency tables with more than two factors it is possible to find several pairs of models that will differ by the same simple parameter and the difference in G^2 will not be the same. This is true because, in fact, the test of significance of a parameter is *conditional* on a specific set of parameters having been included.

Brown (1976) suggested a procedure called *screening* to decide whether or not to include each parameter in the final model. For each *u* term, Brown suggested obtaining two estimates of significance. One estimate, called a "test of marginal association," compares two simple models. For example, in a four-factor table compare [AB] with [A] [B]. A second estimate, called a "test of partial association," compares two complex models; for example, compare [AB] [AC] [AD] [BC] [BD] [CD] with [AC] [AD] [BC] [BD] [CD]. Both differences assess the significance of the AB interaction term, but with different conditionals. The process of comparing the two significance tests of an effect (such as the AB interaction) gives a range of significance values that can be used to decide whether or not to include a particular term in the final model.

Upton (1978) suggested using "forward selection" and "backward elimination." Forward selection means that at each stage of the analysis the most important *u* is included in the model. Backward elimination means that at each stage the least important *u* is excluded from the table. Hocking (1976) pointed out that neither procedure will lead necessarily to a unique best model, if such a model even exists. The goal is to obtain a relatively simple *and interesting* model. Goodman (1971a) wrote:

> By including additional *u*'s in the model, the fit can be improved; and so the researcher must weigh in each particular case the advantages of the improved fit against the disadvantages of having introduced additional parameters in the model. Different researchers will weigh these advantages and disadvantages differently. (p. 41)

10.11 Contrasts

Plackett (1962) proposed that since the estimated variance of the logarithm of a Poisson frequency is the reciprocal of the frequency, that if λ is any weighted sum of the logs of the cell frequencies that is a linear contrast:

$$\lambda = \sum_{i,j,k,\ldots} a_{ijk\ldots} \, ln(m_{ijk\ldots})$$

Then an estimate of the variance of $\hat{\lambda}$ is

$$v(\hat{\lambda}) = \sum_{i,j,k,\dots} (a_{ijk\dots})^2 / \hat{m}_{ijk\dots}$$

Furthermore, Goodman (1971a) showed that the following ratio is asymptotically a unit normal distribution.

$$S(\hat{\lambda}) = \hat{\lambda} / \sqrt{v(\hat{\lambda})}$$

This is a useful method for comparing cells in a table.

10.12 The Analysis of Residuals

One option, once we have found an interesting model that nearly fits, is to examine the table cell by cell and look for cells that may be ruining the fit. To accomplish this analysis of residuals we can use the Freeman-Tukey deviates, or the components of chi square. (See Chapter 3 on statistical inference about Markov chains, section 3.4 on cell-wise examination of contingency tables.) An alternative is to force cells to zero that we believe are "outliers," recompute G^2, and subtract our two G^2 terms; differences in G^2 will have one degree of freedom if only one cell is involved.

10.13 The Homogeneity Problem Again

Most researchers currently studying social interaction are in a descriptive phase of scientific investigation. This means there are coding systems with many code categories. Also, many researchers have relatively short streams of data for each interacting unit (dyad, triad, etc.). With 100 codes there will be $(100)^2 = 10,000$ cells in a first-order Markov timetable. It will thus be difficult to do a sequential analysis unit by unit because there won't usually be enough data. The solution we have been investigating is to pool data across units within each cell of the contextual design.

The problem with this solution, however, is that it assumes there is homogeneity within a cell across units. Fortunately, the assumption can be tested as follows: If units within each cell are randomly divided into two groups, the result will be a saturated design (for each cell) of the timetable crossed by subgroup within cell. The saturated model is compared to one that has only the timetable margins constrained. The LRX^2 test will evaluate whether the subgroup factor or its interactions with timetable variables are significant. If the subgroup factor is denoted factor 1, the k codes at time t by 2, and the k codes at $t + 1$ by 3, we compare the saturated $1 \times 2 \times 3$ model with a model that constrains the 2×3, the 1×3, and the 1×2 margins.

11

LOG-LINEAR MODELS:

REVIEW AND EXAMPLES

11.1 Model Building and Selection of Categorical Data Models

The selection of a log-linear model for describing a given data set becomes increasingly complicated as the number of variables increases, because of the rapid increase in possible associations and interactions. It is normally too cumbersome and demanding a task to fit all possible models once the number of dimensions exceeds three, and we thus need rules of thumb for selecting a specific model over others. As in fitting regression models, we have to balance two contradictory goals or objectives. On the one hand, we would prefer a model complex enough to afford a reasonably good fit of the data. On the other hand, we want a model that is relatively simple to interpret and that is parsimonious – one that smoothes rather than overfits the data.

Even though the partitioning and ordering properties of the likelihood ratio X^2 statistic are very useful in assessing the relative worthiness of several (nested) candidate log-linear models, the process of finding the most appropriate – "best" fitting and parsimonious – (hierarchical) log-linear model is by no means an easy task. In general, we are able to fit all possible models for a three-way table; however, the number of hierarchical models to be fitted increases dramatically as the number of dimensions grows. Thus, for anything larger than a three-way table, we usually have to consider model selection strategies to limit the number to be evaluated for any given problem.

We are unable to cover fully and in detail the intricacies of all the methods proposed for identifying well-fitting unsaturated models., but we shall briefly sketch some of the better-known alternatives.

There are several possible criteria for model selection, dependent on the inferences and interpretations to be made. The more commonly used are parsimony, simple interpretation, all significant effects, and percent of X^2.

Parsimony refers to the choice of a model containing the fewest number of terms and yet yielding a nonsignificant goodness-of-fit.

Simple interpretation may involve a model that has more terms than a parsimonious model but that is less complicated to interpret. For instance, a three-way interaction may often be easier to interpret than a set of three two-way interaction effects.

In a hierarchical model containing *all significant effects*, the changes in the goodness-of-fit (e.g., Goodman's LRX^2 [i.e., G^2] or Pearson's X^2) due to the addition of any single term is nonsignificant, while the change in the test-of-fit owing to the deletion of any single effect is significant. For large multidimensional tables, one may obtain several completely disjoint models that satisfy this criterion of "all significant effects."

When very large sample sizes are involved, tests of all effects are likely to be significant even when the effect is very small, but nonzero. In such a case, an additional possible criterion is the so-called *percent of X^2* relative to the baseline model. One approach is to examine the ratio of likelihood ratio chi-squared test of a model to that of a baseline model, that is, one containing only all possible main effects.

Before proceeding to several model-building strategies, let us reintroduce the notation being used here. We shall write a saturated log-linear model for a four-way table with indices A, B, C, and D as follows:

$$\log \hat{m}_{ijkl} = \hat{u} + (\hat{u}_A)_i + (\hat{u}_B)_j + (\hat{u}_C)_k + (\hat{u}_D)_l$$

$$+ (\hat{u}_{AB})_{ij} + (\hat{u}_{AC})_{ik} + (\hat{u}_{AD})_{il} + (\hat{u}_{BC})_{jk} + (\hat{u}_{BD})_{jl} + (\hat{u}_{CD})_{kl}$$

$$+ (\hat{u}_{ABC})_{ijk} + (\hat{u}_{ABD})_{ijl} + (\hat{u}_{ACD})_{ikl} + (\hat{u}_{BCD})_{jkl}$$

$$+ (\hat{u}_{ABCD})_{ijkl} \quad ,$$

where $\hat{m}_{ijkl}$ is the observed frequency in cell (i,j,k,l) and the $\hat{u}$'s satisfy the usual ANOVA-like constraints of having to sum to zero. An interaction effect is identified only by subscript — e.g., AB depicts the parameters $(u_{AB})_{ij}$ for all i and j.

Because we consider only hierarchical models here, the presence of a Kth order effect ABC... implies the presence of all effects whose factors are subsets of ABC... For instance, the inclusion of AB in the model would imply that A, B, and u (the constant form) are all also in the model; the inclusion of ABC in the model implies that A, B, C, AB, AC, BC, and u are all also in the model; and so forth. Hierarchical models can thus be specified uniquely by the set of highest-order interactions whose presence imply those of the remaining effects. For instance, [AB] [BC] defines the model $u + (u_A)_i + (u_B)_j + (u_C)_k + (u_{AB})_{ij} + (u_{BC})_{jk}$.

The inadequacy of model fit is tested again by the likelihood ratio X^2 statistic (LRX^2), given by

$$LRX^2 = 2\sum_{ijkl} \hat{m}_{ijkl} \log_e \left[\frac{\hat{m}_{jkl}}{m_{ijkl}} \right] ,$$

where $m_{ijkl} = E(x_{ijkl})$ for the model being fit. This test statistic is equivalent to the minimum discrimination information statistic (m.d.i.s.) due to Ku and Kullback (1974).

The process of model building can be divided into two stages: the selection of a starting or baseline model(s), and the addition or deletion of effects from the base model. Possible strategies for choosing a base model include simultaneous tests of order K, Brown's tests of association (see Brown, 1976), Birch's partial association (see Birch, 1964, 1965), and fixed strategy models.

A model of order K is one that contains all K-factor interactions but no higher order forms. For example, in our four-way table, the effects [AB] [AC] [AD] [BC] [BD] [CD] would describe the model of order 2 that contains all main effects and 2-factor interactions.

The difference between the test-of-fit from models of order K and order $K - 1$ provides a test that all the K-factor interactions are simultaneously zero. The maximum such K for which the simultaneous test is significant generally identifies the order at which some (but not necessarily all) of the K-factor effects are needed for an adequate model. It is thus possible to choose the model of order $K - 1$ as a base model, and a stepwise procedure can then be used to add K-factor effects; alternatively, one may delete K-factor effects in a stepwise fashion from a baseline model of order K.

11.2 Brown's Measures of Marginal and Partial Association

As noted, Brown (1976) proposed an approach permitting a screening of terms in a hierarchical log-linear model so that only a limited reasonable number of individual models need be considered in the model-building and model-selection process. For each term in the model, he computed two separate tests, called *marginal* association and *partial* association, to indicate the order of magnitude of changes in goodness-of-fit produced by entering a given effect into the model.

A test of *marginal association* between K factors is defined, the name indicates, as a test that the K-factor interaction is zero in the marginal table formed by these K factors. For instance, if u_{AB} is the effect of interest, the test of marginal association between A and B is obtained by fitting the model [A] [B] to the marginal table indexed by A and B (i.e., the summed table over the levels of the remaining categorical variables of interest). This is equivalent to using the difference between the test-of-fit of [AB] and [A] [B] on the original frequency or contingency table, and is an *unconditional* test of the interaction – namely, it ignores the effect of the other variables.

Brown's test of *partial association* between K factors is defined as the difference in goodness-of-fit of the model of order K and the same model containing all marginals of that order excluding the K-factor interaction of interest. For instance, the partial association between A and B in our four-way table example is the difference between the goodness-of-fit test statistics of [AB] [AC] [AD] [BC] [BD] [CD] and [AC] [AD] [BC] [BD] [CD], that is, u_{AB} is excluded in the latter case. This is thus a *conditional* test of the K-factor interaction adjusted for all other effects of the same order. The degrees of freedom for the test can be easily obtained by subtraction. Since the test is obtained as the difference between nested models, one should note that only the LRX^2 (and not Pearson's X^2) is an appropriate test statistic here.

Brown's tests can be used in several ways to select base models. One very helpful method is to compose an initial baseline model including all effects that are significant according to both the marginal and partial association tests, and to further investigate the addition to such a model of those effects that are significant in only one of the tests of association. Thus, terms get categorized into (a) those that should be included in a baseline model (both tests are significant), (b) those that warrant further examination (only one of the two tests is significant), and

(c) those that need to be excluded from consideration to improve the model fit (both test are nonsignificant).

Birch (1964, 1965) introduced a test of *partial association* between two factors as the test-of-fit of a model with all interactions except those containing both the factors. In our example of the four-way table, one would look at the model [ACD] [BCD] to test the Birch partial association between factors A and B. One would fit this and fit [ACD] [BCD] [AB]. Examination of the test for partial association for all possible pairs of factors leads one to a baseline model that includes all significant second order effects and appropriate higher order relatives, and those which are not also higher order relatives of second order effects whose tests of partial association are nonsignificant. The Brown procedure is available as an option in the BMDP program 4F.

11.3 Standardized Parameter Estimates

In the saturated model, the ratio of the estimates of the parameters (u's) to their corresponding standard errors can be also used to test effects. We may thus fit the saturated model and observe which parameter estimates have relatively larger standardized values. The standardized effects can also be compared to critical values using a simultaneous test procedure to select important effects for a baseline model. A natural starting point for a model is a hierarchical one containing all u-terms that have at least one standardized estimate, $z = \dfrac{\hat{u}}{SE(\hat{u})}$, exceeding in absolute values a certain number, for example, 2.0. These terms will then define a minimum hierarchical model that should be fitted, and additional parameter effects evaluated based on stepwise selection procedures. One good rule of thumb is to simplify the baseline model by eliminating terms and yet maintain an adequate fit to the data, if the baseline model fits well. If, on the other hand, the base model does not fit well, additional parameters having moderate standardized estimates may be added to obtain an adequate fit. One may obviously also use forward selection techniques to arrive at such a model.

11.4 Stepwise Selection Procedures

In a lot of applications the best guide in model selection is the underlying theory behind the research interest. Such baseline models that may be fixed in advance include the choice of the saturated model, the model of all first-order (main) effects, or a model based on prior experience. In a few cases, however, it may be of relevant interest to use some form of an automatic selection method analogous to "forward selection" or "forward stepping" involving stepwise (i.e., addition of effects) and "backward elimination" (i.e., incorporating deletion of effects), procedures employed frequently in multiple regression analysis (see, for instance, Draper & Smith [1981] and Neter, Wasserman, & Kutner [1985]) to help select an appropriate model.

One option is to start with fitting models of a uniform order, seeking a model with terms of order $K - 1$ that fits poorly and a model with terms of order K that fits very well. At such a juncture, one considers models based on configurations of order $K - 1$ and order K, and uses one of the following two approaches:

1.   Forward Selection

     At each step a *forward selection procedure* fits all possible hierarchical models that *include* the current model and differ from it by one effect only; the procedure then *adds* the most significant effect to the model. The process is continued and allowed to terminate when no further terms can be added to significantly improve the model fit.

2.   Backward Elimination

     Here, at each step, a *backward elimination (stepping) procedure* fits all possible hierarchical models that *include* the current model and differ from it by one effect only; the procedure then *deletes* from the model the effect with the least significant change in LRX^2. Other possible criteria for choosing an effect to delete are the smallest change in LRX^2 or change in LRX^2 per degree of freedom. The process is continued step by step and allowed to terminate when no further term of order K can be deleted on the basis of change in goodness-of-fit.

Modifications to these procedures are also available to yield "stepwise forward" or "stepwise backward" procedures (very similar to the case of stepwise regression) such that the inclusion or deletion of a specific effect is not irreversible.

Unfortunately, no single "best" selection strategy exists, since such a choice depends in large part upon our a priori knowledge of the interrelationships among the variables. Without some means of allowing for a "guided search" process, one would have to evaluate an impractically large number of models, with no guarantee of assuring the best selection.

11.5 Examples

11.5.1 Brent and Sykes (1973)

Police-citizen interaction was observed and classified for 1,622 acts. The classification scheme coded acts as definitional, controlling, or resistant and confirming.

Order. A table was formed in which the dimensions were (1) the actions of participant A at time t; (2) B at time $t + 1$; (3) A at time $t + 2$; (4) B at time $t + 3$; and (5) whether A or B was an officer or suspect. The resulting mating was a $3 \times 3 \times 3 \times 3 \times 2$ table. To test for a first-order Markov model the terms u_{12}, u_{23}, and u_{34} represent the associations between acts at time t and $t + 1$ and $t + 2$ and $t + 3$. These terms, when added to the model (see right side of the table) lead to a reduction in G^2 of 527.3, while reducing the degrees of freedom by 12. Hence, these terms contribute significantly to producing a better fitting model (see Table 11.1).

Continuing in this way, the second-order Markov model (adding u_{123} and u_{234}) provides a substantially better fit to the data than a first-order model and does not differ significantly from a third-order model (adding u_{1234}).

Homogeneity and Stationarity. Three similar five-dimensional tables were formed. The first ten and last ten interactions of the encounter were compared to form a time dimension. Four dimensions of the tables were role by the timetable, that is, acts at time t, $t + 1$, $t + 2$. The other two tables crossed their four dimensions by the social class of the suspect, and the nature of the offense. Stationarity was tested by examining the interaction of the time dimension with the model effects (timetable). Homogeneity was tested by examining the interaction of social class and offense by the timetable. Table 11.2 presents the results of analyses of stationarity with the second-order Markov model as well as the two major factors of the contextual design, the nature of the offense and social class. Results show that there is no significant effect of social class or the nature of the offense on the timetable. The stationarity analyses were interesting. To read Table 11.2, find Model 4,

Table 11.1

Model	df	G²	Models	ΔG²	Δdf	P	Terms Differing between Models
							Tests of Order and Contingency
1. $u_1+u_2+u_3+u_4+u_5$	152	3526.3					
2. All $1^*+u_{12}+u_{23}+u_{34}$	140	2999.0	1–2	527.3	12	P < .0001	1st-order other-contingent effects $u_{12}+u_{23}+u_{34}$
3. All $1+u_{12}+u_{23}+u_{34}+u_{13}+u_{24}$	132	2003.9	2–3	995.1	8	P < .0001	2nd-order self-contingent effects $u_{13}+u_{24}$
4. All $1+u_{12}+u_{23}+u_{34}+u_{13}+u_{24}$ $+u_{14}$	128	1984.0	3–4	16.9	4	.001 < P < .01	3rd-order other-contingent effects u_{14}
5. All $1+u_{12}+u_{23}+u_{34}+u_{13}+u_{24}$ $+u_{14}+u_{123}+u_{234}$	112	1878.1	4–5	105.9	16	P < .0001	2nd-order other/self-contingent effects $u_{123}+u_{234}$
6. All $1+u_{12}+u_{23}+u_{34}+u_{13}+u_{24}$ $+u_{14}+u_{123}+u_{234}+u_{124}+u_{134}$	96	1846.7	5–6	31.5	16	P > .01	3rd-order other/other + self/other effects $u_{124}+u_{134}$
7. All $1+u_{12}+u_{23}+u_{13}+u_{24}$ $+u_{14}+u_{123}+u_{234}+u_{124}+u_{134}$ $+u_{1234}$	80	1821.2	6–7	25.5	16	P > .01	3rd-order other/self/other effects u_{1234}

Table 11.2

Model		df	G^2	Models	ΔG^2	Δdf	P	Effects
1. 1,2,3,4,5 *	X = Time	99	1318.9					1. Second-order model
	X = Soc class	99	1995.1					
	X = Offense	99	2424.6					
2. 1,2,3,4,5,23,24,34,234	X = Time	79	940.3	1–2	378.6	20	P<.0001	2. All role effects on second-order model
	X = Soc class	79	1266.9	1–2	728.2	20	P<.0001	
	X = Offense	79	1508.9	1–2	915.7	20	P<.0001	
3. All 1,23,34,24,234,235,245, 345,2345,25,35,45	X = Time	53	133.9	2–3	806.4	26	P<.0001	3. Effects of X on distribution of responses
	X = Soc class	53	38.9	2–3	1228.0	26	P<.0001	
	X = Offense	53	86.2	2–3	1422.8	26	P<.0001	
4. All 1,23,24,34,25,35,45, 12,13,14,234,235,345,245, 2345	X = Time	47	95.6	3–4	38.3	6	P<.0001	4. Effects of X on first- and second-order main effects on distribution of responses
	X = Soc class	47	33.7	3–4	5.2	6	P>.10	
	X = Offense	47	78.3	3–4	7.9	6	P>.10	
5. All 1,23,34,24,25,35,45,12, 13,14,234,235,345,245,123, 124,135,2345	X = Time	35	79.5	4–5	16.2	12	P=.1839	5. Interactions of role and x, and interactions of role, x, and distribution of responses
	X = Soc class	35	27.7	4–5	6.0	12	P=.9134	
	X = Offense	35	62.1	4–5	16.2	12	P=.1824	
6. All 3,2345	X = Time	28	54.3	5–6	25.2	7	P<.001	6. Interaction of role, x, and second-order effects
	X = Soc class	28	25.7	5–6	2.0	7	P>.95	
	X = Offense	28	44.6	5–6	17.5	7	P>.10	
7. All 4	X = Time	8	11.5	6–7	42.8	20	P=.001	Effects of role on distribution of responses
	X = Soc class	8	9.7	6–7	16.0	20	P=.7194	
	X = Offense	8	14.5	6–7	30.1	20	P=.0676	
8. All 1, all 2, u_{123}, u_{234}		104	374.3	5–8	1503.8	8	P < .0001	$u_{13} + u_{23} + u_{35} + u_{345}$ Role x 1st-order other-contingent effects
9. All 1, all 2, u_{123}, u_{234}, u_{135}, u_{235}, u_{345}		92	338.7	8–9	35.6	12	P < .001	$u_{145} + u_{235} + u_{345}$ Role x 2nd-order other-contingent effects
10. All 1, all 2, u_{123}, u_{234}, u_{135}, u_{235}, u_{345}, u_{130}, u_{345}		84	221.3	9–10	117.4	8	P < .0001	$u_{135} + u_{45}$ Role x 2nd-order self-contingent effects
11. All 1, all 2, u_{123}, u_{131}, u_{234}, u_{135}, u_{235}, u_{345}, u_{130}, u_{245}, u_{1230}, u_{2345}		68	98.0	10–11	123.3	16	P < .0001	$u_{1235} + u_{2345}$ Role x 2nd-order other/self-contingent effects

*Where "all 1" means all u's having one subscript (i.e., u_1, u_2, u_3, u_4, and u_5), "all 2" means all u's having 2 subscripts, and so on.

Dimensions 1, 2, 3, 4 = acts at times t, $t-1$, $t-2$, $t-3$

Dimension 5 = role or actor

the first row in which X = time. The models subtracted are 3 and 4, which leave only the 12, 13, and 14 effects in the G^2 term. This represents the stationarity interaction effects with acts at time $t - 1$, $t - 2$, and then $t - 2$ taken separately. The G^2 term is 38.3, with 6 degrees of freedom, which is significant at $p < 0.0001$ and is noted as "effects of X on distributions of responses." Note that it is not the only test of stationarity possible; it represents only one test, namely, the effects of stationarity on the response distribution (either at time t, and $t - 1$, or $t - 2$), not on the timetable itself. While time (first ten versus last ten interactions)

> ... by itself has no significant effects upon the transition probabilities of the interaction process model, time and role together interact to produce significant differences in transition probabilities. (p. 396)

However, Brent and Sykes (1973) argue that these effects are small and suggest that G^2/n is a rough measure of association. For the seven effects for time this coefficient was 42.8/1900 = .02. The equivalent coefficient for the interaction of roles with the Markov model was 1185/1900 = .62. They wrote:

> ... clearly whatever lack of stationarity there may be in this model is considerably smaller than the order of role effects of the model. (p. 397)

For these reasons the data were considered essentially stationary.

A state transition diagram summarizes the role and sequential nature of these police-suspect interactions showing that (see Figure 11.1):

> ... the general flow of the digraph tends to confirm the import of authority and power in the social definition of reality. "Taking charge" perhaps is not as important in the literal behavioral sense, as it is in the cognitive and symbolic sense, for ultimately the encounter turns on who the actors *think*, or who they can be persuaded or coerced into thinking, they are. (p. 401)

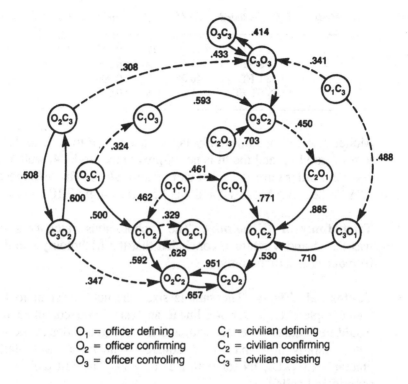

O₁ = officer defining   C₁ = civilian defining
O₂ = officer confirming   C₂ = civilian confirming
O₃ = officer controlling   C₃ = civilian resisting

Figure 11.1. State transition diagram of second-order Markov model with heterogeneous roles (from Brent & Sykes, 1973)

11.5.2 Vuchinich (1984)

Sixty-four recordings (54 video, 10 audio) were made of 52 different families having dinner in their own homes. Tapes were transcribed and coded for "oppositional moves" as either *simple negation* (SN; e.g., "No," "Uh uh," "nah" in the exchange — Jim: I think you've had enough; Mary: No); *disagreement* (DI; Mother: Well it just meets once a week; Daughter: Twice a week); and, *indirect negation* (IN; e.g., Son: I don't want any, I'm full. Father: Always room for Jell-O.) The timetable:

1. **Order.** Vuchinich began by testing order. He denoted time t as F ("first-slot"), time $t + 1$ as S ("second slot"), and time $t + 2$ as T ("third slot"). The table shows his results:

Model	Effect included	LRX^2	df	p
1	FS FT	17.53	12	0.131
2	FS ST	9.40	12	>0.500
3	ST FT	40.06	12	0.000
4	FS ST FT	1.89	8	>0.980

Model 2, which includes only lag-1 transition F to S, S to T, provides a good fit, and the fit is not improved by Model 4, which adds a second order term (FT, first slot to third slot). The difference in LRX^2's between Models 2 and 4 is 7.51, $df = 4$, $p > 0.10$.

2. **Trichotomy or Dichotomy.** The trichotomous category system was tested and Vuchinich concluded that the trichotomy could not be reduced to a dichotomy.

3. **Contextual Effects.** The sample size was not sufficient to form one complete table for log-linear analysis. The complete table would have crossed gender with age for all three codes in the first-order Markov timetable. Each square would be a 3×3 Markov timetable (SN, DI, IN for both t and $t + 1$). The full table would contain 144 cells.

Instead, Vuchinich created a set of smaller tables. One set evaluated the generation effect. These tables were the generation tables, grouped by who initiated. There were two such tables:

	older to younger	older to older
Initiated by older male		
Initiated by older female		

	younger to older	older to younger
Initiated by young male		
Initiated by young female		

Each cell of the tables above contains the timetable. To assess the effects of the gender of the initiator the following two tables were formed:

	female to male	male to female
First slot older		
First slot younger		

	male to female	male to male
First slot older		
First slot younger		

a.   Vuchinich used the ECTA program to fit all possible log-linear models.

b.   For the gender sequence and generation sequence tables, lambdas and their standard deviations were computed. These lambdas are the same as the mu's in our notation.

Table 11.3. Test of fit for selected log-linear models for
tables partitioned by generation and gender of participants
(from Vuchinich, 1984).

Effects included[a]	df	Y–O/Y–Y Table LRX^2	p	O–Y/O–O Table LRX^2	p
FS	27	83.43	0.000	58.02	0.001
GJ	32	123.01	0.000	144.05	0.000
FS GJ	24	32.43	0.117	37.61	0.038
FS SG GJ	22	18.98	>0.5[b]	37.06	0.023
FSG GJ	16	14.11	>0.5	27.27	0.030
FSJ GJ	16	15.60	>0.3	10.60	0.195[b]
FSJ FSG GJ	8	5.31	>0.7	10.79	0.214
		F–M/F–F Table		M–F/M–M Table	
FS	27	43.41	0.024	95.37	0.000
PQ	32	177.22	0.000	96.78	0.000
FS PG	24	28.19	0.252[b]	19.62	0.198
FS SP PQ	22	27.47	0.194	23.95	0.352[b]
FSP PQ	16	19.48	0.216	14.16	>0.5
FSQ PQ	16	18.99	0.269	22.71	0.122
FSP FSQ PQ	8	11.44	0.158	6.62	>0.5

[a]F = First Slot, S = Second Slot, G = Generation Sequence, J = Gender of First
Slot, P = Gender Sequence, Q = Generation of First Slot.
[b]Fits selected as models of choice for analysis.

c. Vuchinich then assessed main effects and interaction effects,
    noting that "different effects are operating on different
    tables" (p. 226) and "the same log-linear model does not fit
    all of the subtables" (p. 226). (See Table 11.3.)

4. **Selected Conclusions.** All analyses involve a search for asym-
    metries.

    a. Simple negation (SN) was less frequent than the other two
        codes: indirect negation (IN) was more frequent than the other
        two codes.

    b. Females are more likely to oppose males than are other males.

c. Fathers are more likely than sons to participate in oppositional interchanges.

d. Sons are more likely than daughters to oppose parents.

e. Children are more likely to oppose mothers than fathers.

f. Parents are more likely to oppose sons than daughters.

g. Children respond to other children with more unmitigated opposition (SN) and less direct opposition (IN) than do parents.

5. **Follow-up Paper.** A recent paper by Vuchinich reported results separately for the initiation of conflict, its attenuation and termination.

a. **Conflict initiation.** An analysis of subsets of these data by Vuchinich (1984) analyzed conflict initiations (see Table 11.4).

Table 11.4 includes only data from dinners where a mother, father, and at least one child were present. Parents initiated somewhat more conflict than children (72 vs. 58 initiations). Parents were as likely as children to be verbally attacked (65 vs. 65), but the father was attacked less often than the mother (26 vs. 39). Children attacked the mother twice as often as they attacked the father.

b. **Conflict attenuation.** Conflicts lasted about the same numbers of conversational turns regardless of who initiated it. Figure 11.2 shows how often conflict sequences of different lengths (from 1 to 12 turns) occurred.

The decline shows an exponential pattern predicted by the assumption that at any point in the conflict the probability of a next conflict turn is a constant. Arundale's programs provide exponential fits to data as a matter of course.

There was a tendency for parent-initiated conflicts to attenuate less rapidly than child-initiated conflicts.

Table 11.4. Conflict initiations by family status
(with mother and father present) [from Vuchinich, 1985]

		To whom				
		Mother	Father	Son	Daughter	
	Mother	X	14	14	8	36
Who	Father	17	X	6	13	36
initiates	Son	12	7	3	6	28
	Daughter	10	5	9	6	30
		39	26	32	33	130

Length and Frequency for Conflict Sequences

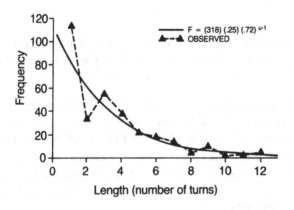

Figure 11.2. Length and frequency for conflict sequences (from Vuchinich, 1984).

c.   **Conflict termination.** Most conflicts (66%) ended in a stand-off in which neither party won or submitted. Some form of submission occurred in 23% of the conflicts and compromise occurred in only 9% of the cases.

Vuchinich's research is interesting because it shows that in normal families conflicts do not get resolved the way therapists train families to resolve them. This suggests that we do not yet understand the prosocial functions of conflict in families. For example, conflict may have the function of showing members of the family

Table 11.5. Allison and Liker's (1982) and Sloane, Notarius, and
Pellegrini's (1985) fit to the data in Table 3.2 (from Sloane et al., 1985)

Model	Marginals fit	*df*	*LRX²*
1	[1] [2] [3]	4	74.69
2	[12] [3]	3	58.66
3	[13] [2]	3	48.45
4	[23] [1]	3	37.02
5	[12] [13]	2	32.41
6	[12] [23]	2	21.00
7	[13] [23]	2	10.78
8	[12] [13] [23]	1	2.23
9	[123]	0	0.00
10	[123']	1	0.26

that there is caring, concern, and involvement. Or perhaps the resolution of these conflicts has a very long time course (perhaps to be
measured in child developmental epochs).

11.6 Nested Hierarchical Models

For a discussion of nested hierarchical models, see Magidson et al.
(1981). We will not describe this procedure here but will provide only
an appetizer, based on one sample. Allison and Liker (1982) critiqued
Gottman's (1979a) analyses comparing the interaction of distressed and
nondistressed couples (see Table 3.2). They fit the data in Table 3.2 with
a series of log-linear models (see Table 11.5). Model 9 is the saturated
model. Model 8 is the Allison and Liker model, which gives expected
frequencies of:

		W_{t+1}			
	H_t	1	0	Odds	Odds ratio
Distressed	1	76 (71.7)	100 (104.3)	.76 (.69)	1.92 (1.61)
	0	79 (83.3)	200 (195.7)	.40 (.43)	
Nondistressed	1	80 (84.3)	63 (58.7)	1.27 (1.44)	1.15 (1.61)
	0	43 (38.7)	39 (43.3)	1.10 (0.89)	

Observed frequencies are not in parentheses in the first two columns, while expected frequencies under Model 8 are in parentheses. The odds ratios for the data are 1.92 and 1.15. This suggests the hypothesis that interaction is more highly patterned in distressed than in nondistressed couples. The expected values from the AL model (Model 8 of Table 11.5) suggest that this hypothesis is incorrect.

However, Sloane et al. (1985) recently fit a nested hierarchical model (see Magidson et al., 1981) to the data. Their model, Model 10 in Table 11.5, fit as well as the AL model; the expected values from this model were:

	H_t	W_{t+1}		Odds	Odds ratio
		1	0		
Distressed	1	75.0	100.0	0.76	1.92
	0	79.0	200.0	0.40	
Nondistressed	1	78.2	64.8	1.21	1.00
	0	44.8	37.2	1.21	

The expected values from this model support the original Gottman (1979a) hypothesis.

This example illustrates how the choice between several equally plausible models from a statistical standpoint can have different scientific implications.

11.7 Worked Example

Krokoff et al. (1988) studied the marital interactions of 52 couples. They classified each speech act in the couples' conversation by the speaker and the positivity of the affect. All data were converted to event sequential data, in which no code can logically follow itself.

11.7.1 Markov Model Omnibus Test

The Arundale (1982) programs SAMPLE and TEST were adapted to perform the order, stationarity, and homogeneity analyses. A series of Anderson and Goodman (1957) likelihood ratio chi-squared tests were conducted to assess whether the Markov model of affect for each couple was zero order (independence model), first order (antecedent/consequent relationship), or second order (dependent on the previous two time units). All 52 models were of first order. The LRX^2 independence test had a mean X^2 of 225.4, $df = 25$; the critical X^2 was 37.7 at an .05 alpha level. No data stream was greater than the first order. These tests had a mean $X^2 = 46.4$, with the critical value equal to 185.3 ($df = 150$) at alpha = .05. The likelihood ratio chi-squared tests for stationarity were based on dividing data streams into thirds (see Gottman, 1979, for the rationale for selecting thirds instead of halves). All 52 first-order models were stationary; the LRX^2 had a mean of 33.8, with the critical value equal to 79.1. Homogeneity analyses on the transition probability matrices conducted on social class had a mean ratio of X^2 to df approximately equal to 3.0; when the data were split by marital satisfaction, the ratio fell to approximately 2.0.

11.7.2 Log-linear Analyses

Table 11.6 is a summary of the frequencies of each cell of the $6 \times 6 \times 2 \times 2$ design. The 6×6 consequent (R) by antecedent (P) affect Markov matrix is computed for each cell of the 2×2 contextual design (white/blue collar [C] and unhappily/happily [S] married). To fit a hierarchical log-linear model to these data we employed (using the BMDP programs; see Dixon, 1985) the commonly recommended screening procedure, in which both partial and marginal associations were computed for each term in the model. Next, both forward selection and backward elimination were employed (see Bendetti & Brown, 1978, and Brown, 1976) to decide on which terms to keep in the model (for details on these procedures, see also Upton, 1978). We added .5 to every cell in the design that was not a logical zero prior to analysis, a procedure recommended by Goodman (1970).

Table 11.7 summarizes the results of the screening procedure. The hierarchical model selected by this procedure would include the terms RPS, RPC, RSC, and PSC, to produce the desired nonsignificant LRX^2 of 23.5 with 25 degrees of freedom. However, because the RPC term was not significant in both the partial and marginal association contribution, it should be dropped from the model. The model then contains an RPS term, an RSC term and a PSC term. The differences in chi square due to adding the RPC term in forward selection and deleting in backward elimination were also not significant ($LRX^2 = 56.6$, $df = 50$). None of the standardized deviates with this reduced model (e.g., Freeman-

Table 11.6. Frequencies of affective interaction patterns
by social class and marital satisfaction

Collar [C]	Marital satisfaction [S]	Antecedent affect (Preaff) [P]	Consequent affect (Postaff) [R]						
			H+	Ho	H−	W+	Wo	W−	Total
White	Unhappy	H+	0	12	0	8	11	12	43
		Ho	8	0	29	33	509	168	747
		H−	2	30	0	3	87	54	176
		W+	7	32	5	0	15	2	61
		Wo	15	506	75	17	0	70	683
		W−	11	162	67	1	64	0	305
	Happy	H+	0	24	2	34	49	15	124
		Ho	23	0	31	96	873	133	1156
		H−	5	36	0	4	43	16	104
		W+	27	79	13	0	41	5	165
		Wo	57	894	40	28	0	24	1043
		W−	11	121	18	4	39	0	193
Blue	Unhappy	H+	0	19	1	11	17	5	53
		Ho	15	0	46	53	520	93	727
		H−	1	41	0	23	142	97	304
		W+	13	43	22	0	26	5	109
		Wo	21	524	134	19	0	62	760
		W−	3	99	100	4	54	0	260
	Happy	H+	0	17	13	30	58	25	143
		Ho	27	0	40	72	729	108	976
		H−	8	50	0	24	123	54	259
		W+	31	70	16	0	48	6	171
		Wo	54	725	146	40	0	61	1026
		W−	24	117	44	7	62	0	254

Tukey deviates) were significant. Hence, the reduced model is satisfactory.

The final model must therefore contain only two types of terms. First are terms stating that preaffect (P) and postaffect (R) change with the interaction of social class (C) and marital satisfaction (S). However, the PSC and the RSC terms (post-affect's and pre-affect's interactions with social class [C] and marital satisfaction [S]) are essentially the same term in our situation because they differ only slightly in edge effects: The pre-affect at time $t + 1$ was the post-affect at time t. Thus, there is just one term to understand here, the RSC term. We need to describe it in detail by reference to the cells of our table.

Table 11.7. Results of screening for best model in log-linear analysis

Effect	df	Partial association LRX^2	Partial association Probability	Marginal association LRX^2	Marginal association Probability
R	5	6564.6	.000		
P	5	6588.5	.000		
S	1	194.4	.000		
C	1	5.9	.015		
RP	25	8891.6	.000	8884.6	.000
RS	5	182.2	.000	203.6	.000
RC	5	135.4	.000	108.3	.000
PS	5	179.6	.000	202.3	.000
PC	5	136.7	.000	109.7	.000
SC	1	0.3	.584	3.6	.057
RPS	25	53.01	.001	46.5	.006
RPC	25	33.1	.129	28.4	.289
RSC	5	43.3	.000	31.0	.000
PSC	5	46.1	.000	31.2	.000

Second, there is a term that notes that the first-order Markov matrix itself associates significantly with only one of our contextual variables, marital satisfaction (S). This term is theoretically important. It means that the truly interactive nature of conflict in marriage (reflected by the Markov matrix) is affected only by marital satisfaction. We also need to describe this RPS term by reference to the cells of our table.

Table 11.8 is the RSC summary table in which the entries are displayed as column percentages. We will proceed by the use of contrasts (see Upton, 1978, p. 63). The only interaction difference in the table that is statistically significant is in the last row, which involves negative affect for wives: White-collar wives, when unhappy (compared to happy) are more negative than blue-collar wives, when unhappy (compared to happy), $z = 3.42$, $p < .001$. This result is consistent with data from survey research (Campbell, Converse, and Rodgers, 1976).

Table 11.8 also summarizes the RS and RC subtables. Contrasts applied to the RS subtable show that unhappily married husbands display more negative affect than happily married husbands ($z = 4.02, p < .0001$), and a similar result holds for wives ($z = 3.79, p < .001$). This replicates a well-known result (e.g, Billings, 1979; Gottman, 1979a; Margolin & Wampold, 1981; Schaap, 1982).

Table 11.8. Subtable of preaffect by social class by marital satisfaction

Affect			White collar		Blue collar	
			Unhappy	Happy	Unhappy	Happy
Husband positive			2.1	4.5	2.4	5.1
Husband neutral			37.1	41.5	32.9	34.5
Husband negative			8.7	3.7	13.7	9.2
Wife positive			3.0	5.9	4.9	6.0
Wife neutral			33.9	37.5	34.3	36.3
Wife negative			15.1	6.9	11.7	9.0
RS	H+	Ho	H−	W+	Wo	W−
Unhappy	.02	.35	.11	.04	.34	.13
Happy	.05	.38	.06	.06	.37	.08
RC						
Blue collar	.04	.34	.11	.06	.35	.10
White collar	.03	.40	.06	.05	.36	.10

Table 11.9. The RPS subtable of postaffect (R) by
preaffect (P) by marital satisfaction (S)

Marital satisfaction (S)	Antecedent affect (P)	Consequent affect (R)						Total
		H+	Ho	H−	W+	Wo	W−	
Unhappy	H+	.00	.32	.01	.20	.29	.18	1.00
	Ho	.02	.00	.05	.06	.75	.12	1.00
	H−	.01	.15	.00	.05	.48	.31	1.00
	W+	.12	.44	.16	.00	.24	.04	1.00
	Wo	.02	.71	.14	.02	.00	.09	1.00
	W−	.02	.46	.30	.01	.21	.00	1.00
Happy	H+	.00	.15	.06	.24	.40	.15	1.00
	Ho	.02	.00	.03	.08	.75	.11	1.00
	H−	.04	.24	.00	.08	.46	.19	1.00
	W+	.17	.44	.09	.00	.26	.03	1.00
	Wo	.05	.78	.09	.03	.00	.04	1.00
	W−	.08	.53	.14	.02	.23	.00	1.00

The RC table shows that blue-collar husbands display more negative affect than white-collar husbands ($z = 4.10$, $p < .0001$), whereas there is no significant class difference in negative affect for wives ($z = .48$, n.s.).

Table 11.9 represents the RPS table, which summarizes the variation with marital satisfaction of the first-order Markov table, collapsing over social class. Cells of interest are the negative affect reciprocity cells ($H- \rightarrow W-$ and $W- \rightarrow H-$). The contrasts for these analyses (happy vs. unhappy) were significant both for $H- \rightarrow W-$, $z = 5.33$, $p < .001$, and for $W- \rightarrow H-$, $z = 6.66$, $p < .001$. As predicted, there was greater negative affect reciprocity for dissatisfied than for satisfied couples.

Our hypothesis was that the couple's "communication orientation" (measured with a self-report questionnaire) would be an important variable in understanding the negative affect reciprocity effect. As previously noted, people who score high on this variable are likely to engage in discussions of disagreements ("engagers"), whereas people who score low on it are likely to avoid such discussions ("avoiders"). We used a median split within each level of marital satisfaction, collapsing across social class, separately for both husband and wife. To assess the effect of this variable, we first estimated its effect on homogeneity, and found that the ratios of LRX^2 to their df's fell from approximately 2.0 to approximately 1.7.

To continue exploring the effects of this conflict avoider/engager variable we selected two pure groups from the sample: those in which both spouses were above the median (engagers) or in which both were below the median (avoiders). We predict that the negative reciprocity effect (difference between happy and unhappy couples) would be greatest for conflict engaging couples. Table 11.10 summarizes the negative affect reciprocity transition probabilities for these two groups. The prediction receives support from this table. Comparing unhappy to happy engagers, for $H- \rightarrow W-$, the z-score for the contrast was 2.96, $p < .003$, and for $W- \rightarrow H-$, the z-score was 2.38, $p < .02$. Note also that negative affect reciprocity is high for happy conflict avoiders. We think this is an artifact of the type of tasks used in this (and other observational) research; requiring couples, who report normally avoiding conflict when they disagree, to resolve a disagreement by discussion is a very difficult situation to handle, even for happily married conflict avoiders. This conflict resolution situation probably has low ecological validity for these couples; they they may not be equipped with the skills engagers have at not reciprocating negative affect with negative affect.

Table 11.10. Transition probabilities as a function of
marital satisfaction and communication orientation

	Transition probabilities	
Groups	H– → W–	W– → H–
Engagers unhappy	.282	.365
Engagers happy	.139	.131
Avoiders unhappy	.222	.218
Avoiders happy	.222	.218

This variable of avoiding/engaging is not assessed in marital interaction research (or therapy) on conflict resolution, and these results suggest that our understanding of the relationship between marital satisfaction and marital interaction might be enhanced by considering this variable.

Appendix 11.1:

Likelihood Ratio Tests

We will be making use of the Wilks theorem, an extremely helpful result in statistics. Its major use here will be in comparing two completely specified models in the case that one model is a special case of the other; the models are called "nested" models. We have one model with df more parameters than the other and two likelihoods L_1 and L_0, then the quantity

$$2 \log_e (L_1 / L_0) \sim X_{df}^2$$

asymptotically.

Let's consider two applications of this result, one for analysis of variance and the other for log-linear models.

(1) **Normal linear models.** In the case of the general linear model in which there are N independent observations from a normal distribution, we have the familiar equations

$$H_0: Y = X_0 \, \beta_o + \varepsilon$$

$$H_1: Y = X_1 \, \beta_1 + \varepsilon \; .$$

Consider the univariate case in which

$$L_o(y) = \frac{1}{(2\pi)^{N/2} \sigma^N} \, e^{-||y-X_o\beta_o||^2/2\sigma^2} \; .$$

If we substitute the maximum likelihood estimate (MLE) of

$$\hat{\sigma}_o^2 = \frac{1}{N} \, || \, Y - X_0 \, \beta_0 \, ||^2 \; ,$$

we obtain

$$L_0(\mathbf{x}, \hat{\sigma}_0) = \frac{1}{(2\pi)^{N/2} \, \hat{\sigma}_0^N} \, e^{-N/2}$$

$$L_1(\mathbf{x}, \hat{\sigma}_1) = \frac{i}{(2\pi)^{N/2} \, \hat{\sigma}_1^N} \, e^{-N/2} \; .$$

The likelihood ratio test is:

$$Is \; \hat{\sigma}_0^N / \hat{\sigma}_1^N > C \; ?$$

Numerator and denominator are not independent, so this does not have an F distribution. We have

Is $\dfrac{(1/N \; SS_0)^{N/2}}{(1/N \; SS_1)^{N/2}} > C$? ,

or,

$$\text{is } \; SS_0 \, / \, SS_1 > C^* \; ?$$

Note that SS_1 is the sum of squares for error for the full model, which is SS_E, or the usual error estimate. The test above is the same as

Is $\dfrac{SS_0 - SS_E}{SS_E} > C^* - 1$?

and

Is $F = \dfrac{(SS_0 - SS_E)/\, v_1}{SS_E / \, v_2} > (C^* - 1) \, v_2 \, / \, v_1$?

which *is* an F test. (Note that $SS_0 - SS_E$ and SS_E, but not SS_0 and SS_E are independent.)

(2) **Multinomial Model.** We have k categories and $X_1, X_2, \ldots, X_k$ are the counts. Model A says the maximum likelihood estimates of the counts are $A_1, A_2, \ldots, A_k$. Model B says they are $B_1, B_1, B_2, \ldots, B_k$. Assume that B is the null hypothesis. The likelihoods are

$$L_A(X_1, \ldots, X_k) = \dfrac{N!}{X_1! \ldots X_k!} \, (A_1 / N)^{X_1} \, (A_2 / N)^{X_2} \ldots (A_k / N)^{X_k}$$

$$L_B(X_1, \ldots, X_k) = \dfrac{N!}{X_1! \ldots X_k!} \, (B_1 / N)^{X_1} \, (B_2 / N)^{X_2} \ldots (B_k / N)^{X_k}$$

$$L_A \, / \, L_B = (A_1 / B_1)^{X_1} \, (A_2 / B_2)^{X_2} \ldots (A_k / B_k)^{X_k}$$

$$\log L_A \, / \, L_B = \sum X_i \log A_i \, / \, B_i$$

If we let A be the saturated model $A_i = X_i = O_i$, the observed count, and $B_i = E_i$ is the expected count under some specified model, then

$$2 \log L_A / L_B = 2 \sum O_i \log O_i / E_i$$

which is G^2.

Appendix 11.2:

Matrix Formulation

Once we have the expected counts under a specific model, such as

$$\log \hat{m}_{ij} = \hat{u} + (\hat{u}_1)_i + (\hat{u}_2)_j + (\hat{u}_{12})_{ij}$$

we can obtain the maximum likelihood estimates. Let

$$\log \hat{m}_{ij} = l_{ij}$$

$$\sum_i \log \hat{m}_{ij} = l_{+j}$$

$$\sum_i \sum_j \log \hat{m}_{ij} = l_{++}$$

I = number levels of first factor

J = number levels of second factor.

Then

$$\hat{u} = l_{++} / IJ$$

$$(\hat{u}_{12})_{ij} = l_{ij} - (\hat{u}_1)_i - (\hat{u}_2)_j - \hat{u}$$

$$(\hat{u}_1)_i = \frac{l_{++}}{J} - \hat{u} \ .$$

This is exactly like a complete factorial analysis of variance.

We can obtain the asymptotic variances of these u terms very easily. Each term can be written in general form using multipliers λ_{ij}. For example, we can write $(\hat{u}_{12})_{ij}$ above as

$$(\hat{u}_{12})_{ij} = \sum \sum \lambda_{ij} \log \hat{m}_{ij} \ .$$

Now $\hat{m}_{ij}$ has variance $N p_{ij}(1 - p_{ij})$. If we denote the variance covariance matrix of the $\hat{m}$'s as Λ, then the diagonal of Λ, $\Lambda_{ii} = N_{p_i}(1 - p_i)$ where p_i is the probability of being in cell i, and

$$\Lambda_{ij} = N p_i p_j \ , \ D_p = \begin{bmatrix} p_1 & & & 0 \\ & p_2 & & \\ 0 & & \ddots & \\ & & & p_N \end{bmatrix}, \ P = \begin{bmatrix} p_1 \\ p_2 \\ \cdot \\ \cdot \\ \cdot \\ p_N \end{bmatrix} \ .$$

If the following matrices are defined, then

$$\Lambda = (D_p - PP\prime)N$$

If we now approximate $\log \hat{m}_{ij}$ by a Taylor series, we obtain

$$\log \hat{m}_{ij} \approx \log m_{ij} + \frac{1}{m_{ij}} (\hat{m}_{ij} - m_{ij}) \ .$$

The $(\hat{m} - m)$ is the only random term; the rest are unknown constants. By arguing in terms of order of magnitude, it is possible to show that various terms go to zero asymptotically so that the matrix of covariances and variances of our contrasts (the MLE of our sums in the model)

$$\sum \lambda_{ij} \left[\log m_{ij} + \frac{1}{m_{ij}} \left[\hat{m}_{ij} - m_{ij} \right] \right]$$

is

$$\left(\frac{\lambda}{m}\right)' \Lambda \left(\frac{\lambda}{m}\right) .$$

If you have a linear contrast in which $\sum \lambda_i = 0$ then the above equation reduces to

$$var \left[\sum \lambda_{ij} \log \hat{m}_{ij}\right] = \sum \sum \lambda_{ij}^2 / m_{ij} .$$

Appendix 11.3:

The Relationship Between

Information Theory and Log-Linear Models

(James T. Ringland)

This section will briefly explore the relationship between information theory and log-linear models. Recall that if we have m alternatives, each of which occurs with probability p_i, the information is

$$H = - \sum_{i=1}^{m} p_i \log_2 p_i ,$$

where the logarithm is to the base 2. Two questions are usually asked. First, what is the average information per symbol in the sequence, and second, what is the *order* of redundancy.

Stepwise Procedure

The first step assumes nothing about the codes except that they are equally likely, so that $H_0 = \log m$. The next step does not make the equal likelihood assumption, so that

$$H_1 = -\sum_{i=1}^{m} p_i \log_2 p_i \ .$$

The next step computes H_2, which is the information in a pair of symbols minus the information in one symbol,

$$H_2 = H(i, i+1) - H_1 \ ,$$

where $H(i, i+1)$ is the information in a pair of symbols. The next step computes H_3, which is the information in a trigram minus the information in a digram,

$$H_3 = H(i, i+1, i+2) - H(i, i+1) \ ,$$

where $H(i, i+1, i+2)$ is the information in triples.

At each step we search for asymmetry in frequencies (of digrams, trigrams, etc.) that is *new* information, in other words, not accounted for by asymmetry in previous orders.

Notation

Assume that there are only two symbols (i.e., codes) in the observation category system. Figure 11.3 shows a chain of m symbols. The one symbol furthest in the past (i.e., the first symbol) will be indexed by the letter j, and the most recent $(m - 1)$ symbols will be indexed by the letter i.

An M-gram involves subscripting with i and j as follows:

$$H(\{M - 1\}gram) = -\sum_{i=1}^{2^{M-1}} p_i \log_2 p_i$$

where: p_i is the probability of an $M - 1$ gram with symbols indexed with i. There are 2^{M-1} of these. If we denote by:

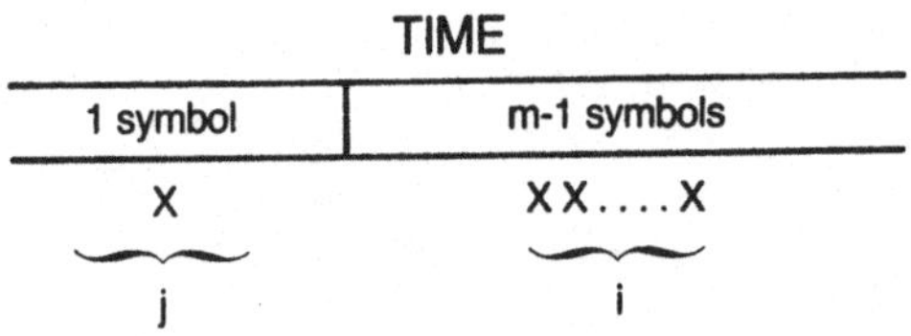

TIME

1 symbol	m-1 symbols
X	X X X
$\underbrace{}$	$\underbrace{}$
j	i

Figure 11.3. Notation for a stream of codes.

p_{ij} = probability of an M-gram, with symbols indexed with (i, j) and

$$p_i = p_{i+} \quad \text{(i.e., summing over } j\text{),}$$
$$= p_{i1} + p_{i2}$$

then

$$H(\{M-1\}gram) = -\sum_i p_{i+} \log_2 p_{i+}$$

and

$$H(M-gram) = -\sum_{i,j} p_{ij} \log_2 p_{ij}$$

But $p_{i+} = p_{i1} + p_{i2}$, so that $H(\{M-1\}gram)$ is

$$H(\{M-1\}\ gram) = -\sum_i p_{i1} \log_2 p_{i+} + p_{i2} \log_2 p_{i+}$$

$$= -\sum_{i,j} p_{ij} \log_2 p_{i+}$$

Therefore H_M, which is $H\{M-gram\} - H\{M-1\ gram\}$ is

$$H_M = - \left\{ \sum_{i,j} p_{ij} \log_2 p_{ij} - \sum_{i,j} p_{ij} \log_2 p_{i+} \right\} \ ,$$

and therefore

$$H_M = -\sum_{i,j} p_{ij} \log_2 \frac{p_{ij}}{p_{i+}} \ .$$

This expression should be vaguely familiar because it resembles a likelihood ratio. However, previously we had counts, not probabilities, so let us transfer to counts. Let n = the total number of M-grams that can be taken out of the observed stream of behavior, and let n_{ij} = the number of M-grams of a specific type. Then

$$\hat{H}_M = - \frac{1}{n} \sum n_{ij} \log_2 \frac{n_{ij}}{n_{i+}}$$

This latter equation illustrates the relationship between information gain and the log-likelihood ratio. The difference between this and the likelihood function is the $1/n$ and the fact that the log is to the base 2. Attneave (1959) multiplied his information measure by n and the $\log_e 2$ to get a chi square, and this is why he used that multiplier.

Appendix 11.4:

Rationale for Comparing Models

Maximum Likelihood: Parameter Estimation

The method of maximum likelihood selects as estimates (called "MLEs") those values that maximize, with respect to the unknown parameters, the joint density of the observed sample. If we have a set of random variables $X_1, X_2, \ldots, X_N$ and a set of observations on these variables $x_1, x_2, \ldots, x_N$, then L, the likelihood of the sample, is the joint probability of $x_1, x_2, \ldots, x_N$. If the $\{X_i\}$ are continuous random variables, then L is the joint density evaluated at the data points. Usually the logarithm of L is maximized. Since the logarithm is an increasing function, it does not matter which is maximized, L or $\log L$.

For example, for the binomial distribution, N trials gives the data $Y_1, Y_2, \ldots, Y_N$ where $Y_i = 1$ if the trial is a success and zero otherwise. To find the MLE for p, the probability of a success, write

$$L = p^y (1-p)^{N-y} ,$$

where

$$y = \sum_{i=1}^{N} y_i .$$

Then

$$\log L = y \log p + (N - y) \log (1 - p) .$$

Using calculus, differentiating L with respect to p and setting it equal to zero, gives

$$\frac{d}{dp} (\log L) = y \frac{1}{p} + (N - y) (\frac{-1}{1-p}) = 0$$

which occurs at

$$\hat{p} = y/N = \frac{1}{N} \sum y_i ,$$

so that the average is the MLE. Note that the second derivative shows this is a maximum.

If the observations are independent and from the same normal distribution, then

$$L = f(x_1, x_2, \ldots, x_N) = f(x_1) f(x_2) \ldots f(x_N)$$

$$L = \prod_{i=1}^{N} \frac{1}{\sqrt{2\pi} \, \sigma} exp \{-(x_i - u)^2 / 2 \sigma^2 \}$$

$$L = \frac{1}{\sigma^N (2\pi)^{N/2}} \, exp \{- \sum_{i=1}^{N} (x_i - u)^2 / 2 \sigma^2 \}$$

Taking the logarithm, differentiating and setting it equal to zero gives:

$$\frac{d}{d\mu} (\log L) = \frac{\sum\limits_{i=1}^{N} (x_i - u)}{\sigma^2} = 0 \ ,$$

which occurs at

$$\hat{u} = \frac{\sum x_i}{N} = \bar{x} \ ,$$

and

$$\frac{d}{d(\sigma^2)} (\log L) = - \frac{N}{2} \frac{1}{\sigma^2} + \frac{\sum\limits_{i=1}^{N} (x_i - u)^2}{2\sigma^4} = 0 \ ,$$

which occurs at

$$\hat{\sigma}^2 = \frac{\sum\limits_{i=1}^{N} (x_i - \bar{x})^2}{N} \ .$$

Thus, for the normal distribution the optimal (maximum likelihood), but not unbiased, estimate of σ^2 is the least squares estimate.

Likelihood Ratio Tests: Comparing Models

Maximum likelihood is used to fit parameters to one model. It does not tell us how to compare two models. The Wilks theorem is again useful. Let us denote by X the observed set of values $(X_1, X_2, \ldots, X_N)$. Assume that the null hypothesis, H_0, gives us a joint density p_0 $(X, \hat{\theta}_0)$, where $\hat{\theta}_0$ are parameter estimates while an alternative hypothesis H_1 gives us a joint density p_1 $(X, \hat{\theta}_1)$, where $\hat{\theta}_1$ are parameter estimates. Then in the specific case where the two models being compared are nested models and $\hat{\theta}_1$ estimates df more parameters than $\hat{\theta}_0$ (i.e., df equal to the difference in the number of parameters in the two models), then (assuming H_0 is true)

$$2 \log (p_1 / p_0)$$

is distributed as a chi-squared distribution with df degrees of freedom. This is an "asymptotic" result, that is, it holds for large N.

12

A SINGLE CASE ANALYSIS OF THE TIMETABLE

Many applications of sequence analysis demand an overall procedure in which the same set of analyses is done for each case (subject, dyad, family, group) in our study. Most of our book is designed with such an application in mind. For example, we will usually want to know, for all the married couples in the study, which one "best" order of the Markov chain to pick, and, in general, how many segments are meaningful for dividing conversations in a stationarity analysis. This is the usual practice, because results would be difficult to report a case at a time.

However, in part to illustrate the flexibility of the methods we have mentioned, we will present a worked everyday example of the analysis of one couple using log-linear methods. We selected a case that has limitations in the amount of data available and that illustrates how much artfulness is needed to select a best model.

This chapter is best thought of as: (1) a worked example of log-linear methods; and as (2) an approach to hypothesis generating with sequential methods.

12.1 A Six-Step Procedure

Remember that the goal is to find the simplest model smaller than the fully saturated model that fits, that is, produces *nonsignificant* chi square. For example, if a Markov (1) (first-order Markov) model produces a significant chi square, this is evidence of the *inadequacy* of the model. The example will also illustrate how to cope with practical problems that can emerge from a model-fitting procedure. As a procedure for converging on the most parsimonious model, we will conduct our analyses in six steps:

Step 1. **Markov Model Analysis.** In this step the approximate order of the largest possible Markov model is decided, by information statistics. It will be decided in part by the number of observations, because sparse tables, i.e., tables with many zeros or a low average cell count, are harder to work with.

Step 2.   **Submodel Fitting.** In this step we search for a smaller model, which should contain, as a submodel, the final model that fits the data well by the usual Pearson or X^2 tests. We are searching for a model smaller than the saturated model with *nonsignificant* chi square. This model fits the saturated table with fewer parameters. There may be problems with this step. For example, because of small cell counts we may have to remove a few cells; we may also have to deal separately with *outlier* cells. This brings us to step 3.

Step 3.   **Residual Examination and Cell Removal.** If a model can be found in step 2 that *nearly* fits, then we examine the standardized residuals (Freeman-Tukey deviates or components of LRX^2) for cells that may be ruining the fit. If there are only a few, we remove these "outlier" cells and return to step 2. We can iterate this procedure if necessary.

Step 4.   **Check All Reasonable Submodels.** In this step we begin with the smallest model and introduce parameters a few at a time to catalog how *each* parameter improves the fit. Conceivably we may be able to simplify the model identified in step 2 by leaving out some terms.

Step 5.   **Check Larger Models.** In this step we add a few new dimensions, representing longer time dependence than the saturated table, and check whether the model previously identified still fits.

Step 6.   **Interpretation of the Model.** In this step we use the *marginal* tables (i.e., the data "smoothed" by the model) to construct probability transition matrices for each of the interactions in the final model. We can simplify each of these by sequentially collapsing (averaging) across the most distant time dependence. (Note, once we select a transition matrix of interest, we can roll the smoothed process forward in time by using the Chapman-Kolmogorov equations; see Chapter 3.)

Table 12.1. Average count per cell for each of several models

Order of Markov model	Size of the saturated model	#Cells	1308 / #Cells = average count per cell
a. Zero	(3×2)	6	218.00
b. First	$(3 \times 2)^2$	36	36.30
c. Second	$(3 \times 2)^3$	216	6.06
d. Third	$(3 \times 2)^4$	1296	1.01

We will now apply these steps to a worked example on the social interaction of a married couple.

12.2 Forming the Timetable

The data for the following example were obtained as follows: A husband and wife met in the laboratory at the end of a day and were videotaped as they discussed the day's events. Their conversation was transcribed verbatim, each utterance serving as a unit of analysis. Using a set of specific cues (see Gottman, 1979a for details) for each speaker (husband or wife), each utterance was coded as either negative (–), neutral (0), or positive (+) affect. Cases in which the husband or wife were both speaking (5 occasions out of 1,308) were omitted. Thus, the codes themselves are in a 3×2 table (affect by speaker), and this 3×2 table was then analyzed for time dependence.

First, Markov models were considered, the Markov (0), the Markov (1), and the Markov (2) models. There is not enough data to go much beyond a Markov (1) model. Table 12.1 shows that the average number of observations (counts) per cell becomes extremely low, i.e., there are many empty cells with the $(3 \times 2)^4$ model. The variables are listed in Table 12.2. The Markov (0) model has no time dependence at all. It is designed to fit only the margins 12, 34, and 56; that is, we do not seek to fit margins that cross time.

The Markov (1) model which has one-step dependence fits the 1234 and 3456 margins. Here we cross one time unit. The Markov (2) model is the saturated one; the expected counts are the observed ones. Even to consider this model, variables 7 and 8 would have to be created (affect and speaker at time minus three), and this would create a $(3 \times 2)^4$ table, with many empty cells. Models that incorporate more time dependence

Table 12.2. Variables in the analysis

#	Variable	Levels
1.	Affect at time zero (0)	−, 0, +
2.	Speaker at time zero (0)	H, W
3.	Affect at time minus one (−1)	−, 0, +
4.	Speaker at time minus one (−1)	H, W
5.	Affect at time minus two (−2)	−, 0, +
6.	Speaker at time minus two (−2)	H, W

Table 12.3. Terms included in the Markov (0) model

Markov (0)		Degrees of freedom
u	Grand mean	1
$(u_1)_i$	Frequencies of behaviors at a fixed time	$2 = (3 - 1)$
$(u_2)_j$	Frequencies of behaviors at a fixed time	$1 = (2 - 1)$
$(u_3)_k$	Frequencies of behaviors at a fixed time	2
$(u_4)_l$	Frequencies of behaviors at a fixed time	1
$(u_5)_m$	Frequencies of behaviors at a fixed time	2
$(u_6)_n$	Frequencies of behaviors at a fixed time	1
$(u_{12})_{ij}$	Interaction in (3×2) behaviors at a fixed time	$2 = (3 - 1)(2 - 1)$
$(u_{34})_{kl}$	Interaction in (3×2) behaviors at a fixed time	2
$(u_{56})_{mn}$	Interaction in (3×2) behaviors at a fixed time	2
Total		16

Table 12.4. Terms included in the Markov (1) model

Markov (1)		Degrees of freedom
Everything in the Markov (0) model		16
$(u_{13})_{ik}$	Relations between time (0) and time (− 1)	4
$(u_{14})_{il}$	Relations between time (0) and time (− 1)	2
(u_{23})	Relations between time (0) and time (− 1)	2
(u_{24})	Relations between time (0) and time (− 1)	1
(u_{35})	Relations between time (− 1) and time (− 2)	4
(u_{36})	Relations between time (− 1) and time (− 2)	2
(u_{45})	Relations between time (− 1) and time (− 2)	2
(u_{46})	Relations between time (− 1) and time (− 2)	1
(u_{123})	Three way time (0), time (− 1) interactions	4
(u_{124})	Three way time (0), time (− 1) interactions	2
(u_{134})	Three way time (0), time (− 1) interactions	4
(u_{234})	Three way time (0), time (− 1) interactions	2
(u_{345})	Three way time (− 1), time (− 2) interactions	4
(u_{346})	Three way time (− 1), time (− 2) interactions	2
(u_{356})	Three way time (− 1), time (− 2) interactions	4
(u_{456})	Three way time (− 1), time (− 2) interactions	2
(u_{1234})	Four way interactions	4
(u_{3456})	Four way interactions	4
Total		66

could be constructed in theory if there were more observations, but that is not feasible in our example.

12.3 Writing the Models Down

The models we will consider are the usual log-linear models, of the form

$$\log (m_{ijklmn}) = u + (u_1)_i + (u_2)_j + \cdots$$

where i indexes variable 1 (affect at time zero), j indexes variable 2 (speaker at time zero), and so on. The terms included in the Markov (0) model are listed in Table 12.3 for the $(3 \times 2)^3$ contingency table. Note that all the interactions (12, 34, 56) are *within* a time; they do not cross times. Terms included in the Markov (1) model are listed in Table 12.4

for the $(3 \times 2)^3$ contingency table. The Markov (2) model would have all of these terms plus time 0, time $- 1$, and time $- 2$ interrelations.

12.4 Preliminary Considerations in Comparing the Models

Recall that the full table (the saturated model) can be compared with any of these models by one of the several X^2 tests, the Pearson X^2

$$X^2 = \sum_{all\ cells} \frac{(O - E)^2}{E}$$

or the likelihood ratio X^2 (LRX^2).

$$G^2 = 2 \sum_{all\ cells} O \log \left(\frac{O}{E}\right)$$

can be compared with a X^2 distribution where the degrees of freedom equal the number of cells minus the degrees of freedom associated with the fitted model.

A few questions can be raised. First, note that, for example, u_{13} and u_{35} should be identical except for edge effects. Because we have omitted 5 points (simultaneous husband and wife speech), our stream of behavior has 12 end points, the starting point, the end point, and, for each of the points removed, the points immediately preceding and immediately following. Recall that u_{13} examines the affect relation between time 0 and time $- 1$ and examines the same affect relation between time $- 1$ and time $- 2$. Thus should approximately equal u_{35} and perhaps these should not be counted twice in the parameter count. On the other hand, they are being estimated (admittedly redundantly) from different cells and the X^2 statistics do have a term from each cell. If we do not count them twice, will our degrees of freedom be too low with respect to the X^2 statistics? Moreover, because of edge effects, especially when small cell counts occur where a difference in one count can be fairly large, are u_{13} and u_{35} actually different? Chatfield (1973) counts them once, while Bishop et al. (1975) count them twice. In the following analysis they have been treated as separate parameters.

Second, how should empty expected cells be handled? Some empty cells represent *sampling zeros*, in which no instances were observed.

Some empty cells represent *structural zeros*, in which logically no instances could occur. If codes cannot follow one another, this would be a structural zero. Parameters related to these structural zeros cannot be estimated. The following adjusted degrees of freedom is proposed by Bishop et al. (1975, p. 115):

> *df* ADJUSTED = (# of cells − # cells with zero expected count) −
> (# of parameters − # of parameters that cannot be estimated)

This is the same as,

> *df* ADJUSTED = original *df* − (# zero expected cells) +
> (# of parameters that cannot be estimated)

12.5 Comparing Models

12.5.1 Step 1: Markov Model Analysis

As noted, because of the sample size we decided to work within the $(3 \times 2)^3$ framework; that is, the data were organized into 216 cell frequencies. In this case, remember, the Markov (2) model is the fully saturated model, which cannot be *tested* unless we make a brief excursion by further subdividing the data. How can we accomplish this? The subdivision $(3 \times 2)^4$ is too fine because the average cell size is 1.01 (see Table 12.1); in other words, there are too many empty cells. We can employ a trick. We can cut the data again either by previous speaker alone, creating a $(3 \times 2)^3 \times 2$ fully saturated model, or by previous affect alone, creating a $(3 \times 2)^3 \times 3$ fully saturated model. These cases have average cell sizes 3.03 and 2.02, respectively. The $(3 \times 2)^3$ fully saturated table was used here to test the Markov (2) model. These Markov models were fit and the log-linear parameters were estimated using the computer program ECTA (which stands for Everyman's Contingency Table Analysis). Since log-linear parameters cannot be estimated when there are zero expected counts in a margin, as there are in the saturated Markov (2) model, a value of .05 was arbitrarily added to each cell. This procedure was recommended by Goodman (1970). The effect of this on the X^2 tests and parameter estimates is unknown, so any results under such a procedure are taken only as somewhat indicative of the structure of the data, but no more than that. All analyses were eventually repeated without the .05 added to each cell.

We need a decision rule to decide when to include parameters in the model. We adopted a conservative rule based on Scheffe's simultaneous

Table 12.5. Terms included in the Markov models

	Significance level		
Model	Weak $.05<p<.10$	Moderate $.01<p<.05$	Strong $p<.01$
Markov (0)		12, 34, 56	1,2,3,4,5,6
Markov (1)	2, 4 1234, 3456	6, 12, 34, 56 124, 346	1,3,5,24,46 13, 35
Markov (2)	6	24, 13, 35	1, 3, 5

comparison procedure. A description of this rule follows. To decide which blocks of parameters (such as the block) to include in the model, we examined the standardized parameters and compared them with $\sqrt{X_{df}^2}$ at $\alpha = .05$, where df = the degrees of freedom associated with the parameter block. This is the limiting form of the Scheffe simultaneous comparison procedure. For example, consider the test of whether all of, say, the u_{13} parameters are simultaneously zero (a X^2 test with 4 df, conducted by comparing 2 models, A and B Model B has all the parameters in A plus the u_{13} parameters. Using the difference of the likelihood ratio X^2 goodness-of-fit statistic, $X_4^2 = G^2(B) - G^2(A)$ will be rejected at level α if and only if *some* combination of the nine cell counts has a standardized value exceeding $\sqrt{X_{\alpha;4}^2}$.

Note that ECTA computes parameter estimates and approximate standardized values for all nine u_{13} parameters $((u_{13})_{--},(u_{13})_{-0},(u_{13})_{0-})$ even though there are only four *independent* parameters. This is true because each row and column is constrained to sum to zero, so that only the upper 2×2 corner of the 3×3 u_{13} table is needed to compute all nine u_{13} parameters. If any *one* of the nine parameters is significant, then we can conclude that the whole batch of parameters is significant, because they are all closely related.

Table 12.5 shows the collections of terms found to be significant. Everything in the Markov (0) model is significant. *Note that this means that the Markov (0) model provides a poor fit to the fully saturated model.* Almost everything in the Markov (1) model is also significant, but the amount of significance is reduced in the Markov (2) full model. This means that the final model we can fit will probably be somewhere

Table 12.6. Pearson X^2 and likelihood ratio LRX^2

Margins fit	Pearson X^2	LRX^2	df
1234 3456 135	161.62 (good fit)	175.69*	138
1234 3456 246	203.08**	209.62**	148
1234 3456 135 246	152.33 (good fit)	168.61*	136

* $p < .05$
** $p < .01$

between the Markov (1) and the Markov (2) models. To find the final fit we will have to be very careful; for example, we will look at the goodness-of-fit chi squares and get rid of the arbitrary .05 that we added on to each cell.

12.5.2 Step 2: Submodel Fitting

We begin with the Markov (1) model, which fit 1234 and 3456 margins, and begin adding some Markov (2) terms. Recall that we are looking for *nonsignificant* X^2 because this means that the model is not different from the fully saturated case. See Table 12.6. Do the first and last models fit the data? The Pearson statistic would lead us to conclude that they do, whereas the *LR* statistic would lead us to conclude that they do not. To examine the situation, we go to the third step, examining the residuals.

12.5.3 Step 3: Residual Examination and Cell Removal

We need some way of standardizing the residuals. Unlike regression analysis, the residual terms, which are the Observed minus Expected terms, are not directly comparable to one another. This is because when the observed count is large we can reasonably expect the residual to be somewhat larger than when the observed count is small. A difference of 10 is not shocking when the observed count is 150, but it is if the observed count is 2 and we expected 12. As indicated in the section on individual cell-wise examination, there are three alternative methods for standardizing the residuals (see Bishop et al., 1975, section 4.4.1): (1) Examine the components of the Pearson X^2, that is, examine $(Obs - Exp)/\sqrt{Exp}$; (2) examine the components of the LRX^2, that is, examine $2\,Obs \log\,(Obs\,/\,Exp)$; or (3) examine the Freeman-Tukey deviates, that is, examine $\sqrt{Obs} + \sqrt{Obs + 1} - \sqrt{4\,Exp + 1}$. All three will be very close if the *Obs* and *Exp* counts are fairly large. However, the FT deviates are more stable when expected counts are small, as they will be in a sparse table.

For the marital interaction data, the largest FT deviate occurred in the cell: Variable 1 (affect at time 0) = level 1 = (−); Variable 2 (speaker at time 0) = level 1 = Husband; Variable 3 (affect at time −1) = level 2 = (0); Variable 4 (speaker at time −1) = level 1 = Husband; Variable 5 (affect at time −2) = level 1 = (−); Variable 6 (speaker at time −2) = level 1 = Husband. This cell, which represents the sequence $H - H0 \rightarrow H-$, had an expected count of 4.20, but *none* were observed. It also had an FT deviate of 3.22, which is somewhat large, although not unreasonably large among 216 observations.[1] Since no other cell had an unreasonably large FT deviate, the next step was to return to Step 2 to assess what would occur if this cell were to be considered an outlier. When testing the maximum residual, we are really testing all 216 to find the largest, so we must be cautious when evaluating its significance.

The expected value of this cell was set to zero. Expected cell counts were not very different, but the X^2 values changed. The Pearson X^2 was 145.24 and the LRX^2 was 158.69, with $df = 135$ (= 136 − 1 cell removed). Both were indicative of a good fit. This is thus a good candidate for the final model, but it must be tested against smaller submodels and larger models.

12.5.4 Step 4: Check All Reasonable Submodels

Several potential improvements are available. Something less than a full Markov (1) model, to which we have added the 135 and 246 margins, could provide an adequate and simpler model than the Markov (1) model with the 135 and 246 margins. Thus, it is first worthwhile examining the

---

[1] $P(\,|Normal|>3.22) = .00126 = 27/216$.

structure of the Markov (1) part of the model. Second, we can ascertain if there are other models near the one we have selected above that would remove the need to set the one cell expected count to zero by fiat.

A whole string of nested models were fitted with and without the one cell removed. The degrees of freedom were adjusted for zero-expected cells. The results are summarized in Tables 12.7 and 12.8. The Pearson X^2 and LRX^2 goodness-of-fit statistics are of interest for the last four models. Model 2 with one cell removed is the smallest model in which everything fits at the .05 level.

The conclusion is that, in a $(3 \times 2)^3$ context, we need all the Markov (1) terms because all are significant except u_{14}, u_{36}, u_{23}, and u_{45}; however, we are compelled to include them because there are important three-way interactions that include these terms.

It is worthwhile to pause and investigate what the 1234, 3456, 135, 246 model means. It implies that the current affect depends on the affect two steps back and that the current speaker depends on the speaker two steps back in a way not adequately described by the Markov (1) model. However, the relation between speaker and affect is only carried over one time unit. There is an exception. The $H - H0 \rightarrow H-$ triple occurs less often than this model might predict. This is what we have learned so far from our first four steps.

12.5.5 Step 5: Check Larger Models

We have identified as a submodel of the full Markov (2) model one that involves two kinds of two-step time dependence: (a) the affects at time −1 and −2 are related to the affect at time 0; and (b) the speakers at times −1 and −2 are related to the speaker at time 0. We now need to check that affect at time −3 is not really needed in (a) and that the speaker at time −3 is not really needed in (b).

To check the first possibility about affect, we constructed a $(3 \times 2)^3 \times 3$ table with the added dimension: Dimension 7 = affect at time −3. We checked the significance of a third order term by comparing the following two models:

Model 1:   The same model as before, in the larger context margins are
             1234, 3456, 567, 135, 357, 246.

Model 2:   A larger model that adds u_{17}, u_{137} u_{157}, u_{1357}, that relates affect at time -3 to affect at time 0. Margins are 1234, 3456, 567, 1357, 246.

The results yielded LRX^2 (Model 1) = 517.95, with df = 544, and LRX^2 (Model 2) = 489.88, df = 501. Once again the df's were adjusted for zero expected counts. The difference between these two X^2 was the contribution of the added terms to a better fit. This difference was 28.07 with df = 43. This is not significant. There is thus insufficient evidence that the added terms contribute to a better fit.

To check the second possibility about speaker, we constructed a $(3 \times 2)^3 \times 2$ table, with added dimension: Dimension 8 = speaker at time -3. There are actually only seven dimensions in this table, but for clarity of presentation the numbers 1, 2, 3, 4, 5, 6, 8 are used, so that even numbers correspond to speaker dimensions. We now compared:

Model 1:   Second-order model. Margins are 1234, 3456, 568, 135, 246, 468.

Model 2:   Add speaker (-3) to speaker (0) relation parameters u_{248}, u_{268}, u_{2468}. Margins are 1234, 3456, 568, 135, 2468.

The results were that LRX^2 (Model 1) = 349.78, with df = 344, and LRX^2 (Model 2) = 342.98, with df = 340. The difference was 6.80, with df = 4. There is thus insufficient evidence of a significant third-order speaker interaction.

12.5.6 Step 6: Interpretation of the Model

Now that we have a model that fits, let us examine it. The shape of it reads: (a) affect depends on affect one and two units in the past; (b) the speaker depends on the speaker one and two units in the past; (c) any relation between speaker and affect only extends one step back; except that, (d) the $H - H 0 H -$ triple occurs less frequently than would be expected in the above model. To examine the dependencies in (a), (b), and (c) above, we can examine the margins fit in the final model and reduce these to transition probabilities. The smoothed (i.e., model) transition probabilities are the expected counts divided by the row sum. For example, the affect dependency is described by the 135 margin.

Table 12.7. Nested models — all cells considered

	Margins fitted	Log-linear model terms	Degrees of freedom Total	Change	LRX^2 Value Total	Reduced
Markov (0)	1 2 3 4 5 6	u u_1 u_2 u_3 u_4 u_5 u_6	206		695.22	
Model	12 34 56	Add u_{12} u_{34} u_{56}	200	6	626.39	68.83
	12 34 56 13 35	Add u_{13} u_{35}	192	8	438.68	187.71
	12 34 56 13 35 24 46	Add u_{24} u_{46}	190	2	413.15	25.53
	12 34 56 13 35 24 46 14 36	Add u_{14} u_{36}	186	4	407.82	5.33**
	124 346 56 13 35	Add u_{124} u_{346}	182	4	315.43	92.39
	124 346 56 13 35 23 45	Add u_{23} u_{45}	178	4	312.10	3.33**
	124 346 56 123 345	Add u_{123} u_{145}	170	8	284.41	27.69

		1234 3456	Add u_{134} u_{356} u_{234} u_{456} u_{1234} u_{3456}	150	20	216.55	67.86
Markov (1)	(1)	1234 3456	Add u_{135}	138	12	175.69	40.86
Tentatively	(2)	1234 3456 135 246	Add u_{246}	136	2	168.61	7.08
Identified	(3)	1234 3456 1235 246	Add u_{1235}	115+	21+	142.07	26.54**
	(4)	1234 3456 1235 1246	Add u_{1246}	107+	8+	137.06	5.01**

** not significant. Significant effect = no asterisk.

Table 12.8. Nested models — one cell omitted

	Margins fitted	Log-linear model terms	Degrees of freedom		LRX^2 Value	
			Total	Change	Total	Reduced
Markov (0)	1 2 3 4 5 6	u u_1 u_2 u_3 u_4 u_5 u_6	205		676.97	
Model	12 34 56	Add u_{12} u_{34} u_{56}	199	6	610.74	66.23
	12 34 56 13 35	Add u_{13} u_{35}	191	8	426.78	183.96
	12 34 56 13 35 24 46	Add u_{24} u_{46}	189	2	403.98	22.80
	12 34 56 13 35 24 46 14 36	Add u_{14} u_{36}	185	4	398.98	5.41**
	124 346 56 13 35	Add u_{124} u_{346}	181	4	307.81	90.76
	124 346 56 13 35 23 45	Add u_{23} u_{45}	177	4	304.57	3.24**
	124 346 56 123 345	Add u_{123} u_{345}	169	8	275.67	28.90

Markov (1)	1234 3456	Add u_{134} u_{356} u_{234} u_{456} u_{1234} u_{3456}	149	20	208.80	66.87
	1234 3456 135	Add u_{135}	137	12	166.73	42.07
Tentatively	1234 3456 135 246	Add u_{246}	135	2	158.69*	8.04
Identified	1234 3456 1235 246	Add u_{1235}	114+	21+	131.78**	26.91**
	1234 3456 1235 1246	Add u_{1246}	106+	8+	127.21*	4.50**

* $.05 < p < .1$ ** not significant   + *df* adjusted   Significant effect = no asterisk.

Table 12.9. The affect relation specified by the model

Past affect		Present affect		
Time −1	Time −2	−	0	+
−	−	125	88	22
−	0	94	116	10
−	+	15	15	9
0	−	81	123	12
0	0	126	230	31
0	+	13	32	9
+	−	17	15	8
+	0	8	18	25
+	+	13	21	16
−	−	.53	.38	.09
−	0	.43	.53	.05
−	+	.38	.38	.23
0	−	.38	.57	.06
0	0	.33	.59	.08
0	+	.24	.59	.17
+	−	.43	.38	.20
+	0	.16	.35	.49
+	+	.26	.42	.32

Table 12.9 presents these smoothed (i.e., model) counts and transition probabilities.

We can sum across the time −2 dimension to obtain the 2×2 table of counts and transition probabilities presented in Table 12.10, but we lose important information in this process.

Finally, we can collapse across the past to obtain the counts 490, 658, 143, and unconditional probabilities .38, .51, and .11, for −, 0, and + affect, respectively.

The speaker dependence is examined via the 246 margin (see Table 12.11). This table suggests that the husband is more likely than the wife to yield the conversational turn (or that the wife is more likely to hold on to it). Collapsing across time −2 gives Table 12.12. A table like it was used by Strodtbeck (1951) as a measure of dominance. Collapsing across time gives counts 585 and 706 and unconditional probabilities .45 and .55 for the husband and the wife as speakers, respectively.

Table 12.10. The affect relation, collapsing over time −2

Past affect	Present affect		
	−	0	+
−	234	216	40
0	220	387	51
+	39	54	50
−	.48	.44	.08
0	.33	.59	.08
+	.27	.38	.35

Table 12.11. The speaker dependence specified by the model

Past speaker		Present speaker	
Time −1	Time −2	H	W
H	H	103	132
H	W	131	220
W	H	187	161
W	W	164	192
H	H	.44	.56
H	W	.37	.63
W	H	.54	.46
W	W	.46	.54

Table 12.12. One-step speaker dependence specified by the model

Past speaker (Time −1)	Present speaker	
	H	W
H	234	352
W	351	354
H	.40	.60
W	.50	.50

Table 12.13. The one-time dependence of speaker-affect interaction

Past	Present					
	H−	H0	H+	W−	W0	W+
H−	25	49	3	68	36	11
H0	24	106	8	96	82	25
H+	4	7	8	15	7	12
W−	59	66	18	82	68	8
W0	64	85	9	36	112	10
W+	15	28	7	4	12	22
H−	.13	.26	.02	.35	.19	.06
H0	.07	.31	.02	.28	.24	.07
H+	.08	.13	.15	.28	.13	.23
W−	.20	.22	.06	.27	.23	.03
W0	.20	.27	.03	.11	.35	.03
W+	.17	.32	.08	.05	.14	.25
Unconditional Probability	.15	.26	.04	.23	.25	.07

To examine the one-time dependence of the speaker-affect interaction, we can look at either the 1234 or the 3456 margin. They are nearly identical. Table 12.13 presents the 1234 margin. We can see that if the husband's affect is negative, then the most likely next event is $W-$. Since $p(W- \mid H-) = .35$; $p(W-) = .23$, we use the fact that the variance of $p(W-)$, under the null hypothesis of no relationship between $H-$ and $W-$ is $p(W-) (1 - p(W-))/N$, where N is the frequency of joint occurrence of $H-$ and $W-$ ($N = 68$ here). Then the binomial z-statistic is asymptotically distributed as $N(0, 1)$.

In this case $z = 2.73$, which is significant. Similarly, the z-score for the conditional probability $p(W- \mid W-)$ is .94, n.s. The wife is thus not likely to continue in a negative affect state. The interpretations of this model can be used to generate descriptive hypotheses that can then be tested for generality across married couples in the study.

Appendix 12.1

Freeman-Tukey Deviates

Some of the result can be roughly derived as follows. Assume the Poisson sampling model. In that case $(Observed)_i = Obs$ is distributed as Poisson (λ), where λ is the expected count. For large samples, Poisson (λ) is nearly distributed as Normal (λ, λ). Using a Taylor expansion,

$$\sqrt{x+a} \cong \sqrt{a} + \frac{1}{2}\frac{x}{\sqrt{a}},$$

we have

$$\sqrt{Obs} = \sqrt{Exp + (Obs - Exp)} = \sqrt{Exp} + \frac{1}{2}\frac{(Obs - Exp)}{\sqrt{Exp}}.$$

Thus

$$\sqrt{Obs} \cong \sqrt{Exp} + \frac{1}{2}\frac{1}{\sqrt{Exp}}(N(0, Exp)) \cong \sqrt{Exp} + N(0, \frac{1}{4}) \sim N(\sqrt{Exp}, \frac{1}{4}).$$

Thus we have the approximation

$$\sqrt{Obs} \sim N(\sqrt{Exp}, \frac{1}{4}).$$

Hence,

$$\sqrt{Obs} \sim Normal\,(\sqrt{\lambda}, \frac{1}{4}),$$

where $\lambda = Exp$. If Obs is large,

$$\sqrt{Obs} + \sqrt{Obs+1} - \sqrt{4Exp + 1}$$

is approximately equal to

$$2 \sqrt{Obs} - 2 \sqrt{Exp} = 2 (\sqrt{Obs} - \sqrt{Exp}) ,$$

which is distributed as

$$2(N(0, \frac{1}{4}) = N(0, 1) .$$

13

LOGIT MODELS

AND LOGISTIC REGRESSION

13.1 Introduction

Logit models are a class of models used to explore the relationship of a dichotomous dependent variable to one or more independent variables. In these models, the logit, or log-odds (i.e., the natural logarithm of the odds), that the dependent variable has a specific given value is analyzed as a linear function of the independent variables. Logit models are analogous to ordinary regression models in which the expected value of a continuous dependent variable is expressed as a linear combination/function of one or more independent variables, in much the same way as hierarchical log-linear models have been earlier shown to be analogs of the analysis of variance (ANOVA) class of models. Generally utilized algorithms for the estimation of logit models exploit such analogy.

As social scientists we are often concerned with the problem of explaining and predicting behavior. Often, the dependent variable that describes behavior is a continuous variable. In that case we can employ standard parametric inferential procedures (like multiple regression analysis), which allow inferences about "average" population behavior given a random sample of data from a population of individuals.

In most observational coding systems, however, the dependent variable is not continuous, but instead is a set of alternatives that are discrete or "quantal." Efforts to analyze behavior observed as discrete outcomes or events thus involve a class of models with discrete, or qualitative, dependent variables. Such models are generally referred to in the social science literature as "quantal choice models" or "quantal response analysis." Examples would include efforts at modeling voting behavior, occupation choice, marital interaction, and so on.

Statistical analysis of general population choice behavior is complicated because such behavior has to be described in probabilistic terms. Thus, quantal response models attempt to relate the conditional probability of a particular choice being made to various explanatory factors including attributes of the alternatives as well as characteristics of the decision makers.

The simplest form of such models is the so-called binary choice model, that is, where the choice is essentially a "yes" or a "no" (represented by a 1 and a zero). A familiar example is an observational category system with two categories, a "positive" versus a "negative" response in the case of marital interaction. Here, one is interested in a way to assess the effects of exogenous explanatory or "independent" variables on a dichotomous "response" or "dependent" variable.

Standard parametric techniques like regression and analysis of covariance, and the like, however, are unfortunately not well suited for analysis of such data. Linear probability models (models that relate the probability of an event or a choice to a series of exogenous factors in a linear fashion) have often been used for this purpose. However, such models are most often unrealistic, and standard estimation procedures based on ordinary least squares (OLS) generally lead to biased and inconsistent estimates. In our discussion, we will: (1) point out the difficulties in applying usual regression techniques to such models; (2) summarize proposed solutions that may be deemed applicable under alternative circumstances; and (3) note some advantages and disadvantages of the methods proposed under (2).

To recapitulate, we indicated our interest in investigating the dependence of a qualitative response on one or more continuous (i.e., internal or ratio-scaled), ordered (or ordinal), or nominal independent variables. The study of this problem has a long history. This type of problem was originally investigated by Bliss (1935) and Fisher (1935) in connection with biological assay. They developed the method of "probit analysis," in which they assumed that the problem involved a binomial response and one continuous independent variable. Later, Berkson (1944) developed logit analysis for the same problem. With the advent of the high-speed digital computer and efficient programming algorithms, more complex experimental situations have been examined by probit and logit analyses and other related methods. The case of binomial response to several independent variables was initially considered by Walker and Duncan (1967) and Theil (1970); multinomial response problems, on the other hand, were examined by Bock (1975), Theil (1970), and Haberman

(1974, 1979). Logit models for factorial tables have been explored by Dyke and Patterson (1952), Bishop (1969), Haberman (1978), and Fienberg (1980). This is the case where one wishes to analyze the effects of categorical variables on a dichotomies or binary response variable, treating all other variables as "design" or "experimental" variables (or factors) that one wishes to manipulate.

As pointed out by Haberman (1974, 1978) and Fienberg (1980), logit models contain terms corresponding to those in log-linear models. However, they deserve special treatment among the class of log-linear models, since the algorithm for logits is simpler than that for log-linear estimation and since interpretations of parameter estimates in logit models are more clearly analogous to interpretations in the case of ordinary regression models. Some researchers fit logit models to datasets in which the remaining nonresponse or independent variables are not design variables, and thus they may lose information on the relationships among the nonresponse variables (see Bishop, 1969). If, on the other hand, our interest lies only in the effects of the other explanatory variables on a given response variable, then an analysis based on logits for the response variable is not only entirely appropriate but also a more efficient estimation procedure. Moreover, *unlike log-linear models, logit analysis can be applied whether the independent variables are discrete or continuous*; however, the methods involved and the estimation algorithms are not necessarily the same. *Also, the simple logit models can be extended to cover the situation where the response variable is not dichotomous but rather "polychotomous" (generally referred to as "polytomous")* via the so-called multinomial choice models and/or conditional logit models due to Nerlove and Press (1973) and McFadden (1973), respectively. This is the case when the codes of an observational system are considered the dependent variable, and thus is the most general case for analyzing observational data when we wish to use continuous independent variables.

13.2 Why We Need the Logit Transformation

13.2.1 The Difficulties of the Standard Regression Model: "Linear Probability Model"

The difficulties of using standard regression procedures when a discrete choice model is adopted will first be illustrated through a simple example. Suppose we are interested in factors that determine whether a husband is nasty or nice to his wife when he returns home after a day's work. We can represent the husband's behavior by using a dummy or indicator variable, which takes on the value of 1 or zero depending on whether the husband is nasty or nice, respectively. Assuming also that such a decision depends on income and random effects, the model can be represented as follows:

$$y_i = \alpha + \beta x_i + e_i, \quad i = 1, \ldots, M$$

where y_i is a random variable that takes the value 1 if the ith husband is nasty, x_i is the ith husband's income, e_i is a random disturbance or error term, and α and β are unknown parameters to be estimated.

Such a model specification has several obvious problems. First, if one makes the usual assumption that the random variable e_i has mean zero, that is, $E(e_i) = 0$ for all i, one must face the fact that although y_i can take only two values (namely, zero and 1), the systematic or nonrandom portion of the right-hand side of the above equation can take any value. Thus, e_i can take on only two values given x_i, namely $-(\alpha+\beta x_i)$ and $1 - (\alpha+\beta x_i)$. Furthermore, if $E(e_i)$ is to be zero, e_i can possibly take these two values with fixed probabilities of $1 - (\alpha+\beta x_i)$ and $(\alpha+\beta x_i)$, respectively. Since $(\alpha+\beta x_i)$ can take on values greater than 1 or less than zero, our position becomes clearly awkward and embarrassing when we attempt to apply a probability interpretation to the model, since the predicted probabilities outside of the [0,1] interval have no meaning. Secondly, given these results, the usual assumption that $E(e_i^2) = \sigma^2$ is no longer valid. The Bernoulli character of y_i really implies a variance of the e_i of $(\alpha+\beta x_i)(1-\alpha-\beta x_i)$; since this depends on the i, the e_i are heteroskedastic, that is, the error terms have unequal variance. Hence, apart from the earlier problem, the use of ordinary least squares (OLS) will result in inefficient estimates and imprecise predictions. Third, the fitted relationship is extremely sensitive to the values taken by the explanatory variable(s), especially where they are "bunched" or concentrated together. Fourth, the usual tests of significance (t-tests) for the estimated coefficients do not apply, the estimated standard errors are not consistent (that is, the estimates are not exactly equal to the parameters, when the entire population is observed), and the R^2 measure is not meaningful. Moreover, since the y_i's are not normally distributed, no method of estimation that is a linear combination of the x_i's is, in general, efficient; in other words, it is theoretically possible to improve on any least squares estimator. In brief, we need a transformation that will overcome these obstacles.

13.2.2 Transformation Approaches

All the difficulties associated with the linear probability model, as outlined in the previous sections, point to the need for alternative model specifications. Given that the most serious set of problems arises from the fact that predictions may lie outside the [0,1] interval, it is only natural to search for alternative distributional assumptions for which all predictions must lie within the appropriate interval. The obvious solution is to transform the original model in such a way as to have all predictions lie in the [0,1] interval for all x_i. Because our primary concern is to interpret the "dependent" variable, y_i, in our model as the probability of making a choice, for example, the husband's choice to be nasty or nice to his wife, it is only reasonable to utilize some notion of probability as the basis for the requisite transformation.

We want the kind of transformation that translates the values of the continuous attributes x_i (the husband's income, in our example) to a probability of the husband's being nasty or nice, which ranges in value from zero to 1. We would also like the transformation to be monotonic, that is, for it to maintain the property that increases in the vector of x_i variables are associated with increases (or decreases) in the dependent variable for all values of the independent variables.

These requirements lead directly to the use of a cumulative probability function or cumulative distribution function (CDF) as a suitable transformation. The CDF is defined as having as its value the probability that an observed value of a variable X (for every X) will be less than or equal to a particular X. The range of the CDF is the [0,1] interval, since all probabilities lie between zero and 1.

The resulting probability distribution may be represented as

$$P_i \equiv \text{Prob}\,(y_i = 1) = F(\alpha + \beta x_i) = F(z_i)$$

and

$$1 - P_i \equiv \text{Prob}\,(y_i = 0) = 1 - F(\alpha + \beta x_i) = 1 - F(z_i)\ ,$$

where F denotes the cumulative density function, and x is stochastic in nature. It is worth noting here that under the assumption that one transfers the model using a cumulative *uniform* probability function, one gets the constrained form of the linear probability model, namely, $P_i = \alpha + \beta x_i$. Although several alternative CDFs can be suggested as worthy candidates for the transformation we seek, we will consider only two, namely, the ones based on the *normal* and the *logistic* distributions. These alternative functional forms are suggested for the dichotomous case, and we will later extend them to the polytomous situation.

13.2.3 Probit Transformation

The *probit probability model* is associated with the cumulative normal probability function.[1] The definition of the cumulative normal distribution gives

$$P_i = F(z_i) = \frac{1}{\sqrt{2\pi}} \int_{-\infty}^{z_i} e^{u^2/2} du$$

where u is a random variable distributed as a standard normal deviate, that is, $u \sim N(0,1)$, or that u is normally distributed with mean zero and unit variance. By construction, the variable P_i will lie in the [0,1] interval; P_i represents the probability of an event's occurring. Since this probability is measured by the area under the standard normal curve between $-\infty$ and $+z_i$, the event will be more likely to occur the larger the value of the index $z_i = \alpha + \beta x_i$. If the parameters of the model were both negative, this could imply that blue-collar husbands are more likely to be nasty to their wives than white-collar husbands; the reverse could be modeled with positive parameters (alpha and beta) values.

For readers unfamiliar with the cumulative normal function, we present Table 13.1, in which the basic relationship is described for specific values of z.[2] The cumulative normal function will later be graphically depicted in comparison to the cumulative logistic function in Figure 13.1, after we have discussed the logistic function.

[1]For further details, see D. J. Finney, *Probit Analysis*, 3rd edition (Cambridge: Cambridge University Press, 1971).

[2]The z's are usually normalized to have unit variance. A slightly different normalization leads to the so-called normit model; this has no substantive impact on the overall discussion here.

Table 13.1. Values of the cumulative normal function

Z	F(Z)
− 3.0	.0013
− 2.5	.0062
− 2	.0228
− 1.5	.0668
− 1	.1587
− .5	.3085
0	.5000
.5	.6915
1	.8413
1.5	.9332
2	.9772
2.5	.9938
3.0	.9987
3.5	.9989

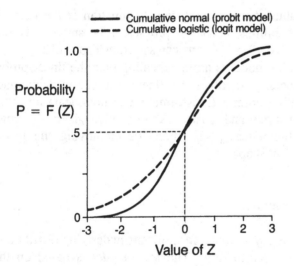

Figure 13.1

To obtain an estimate of the index z_i, one needs to apply the inverse of the cumulative normal function; thus

$$Z_i = F^{-1}(P_i) = \alpha + \beta x_i \ .$$

We can interpret P_i resulting from the probit model as an estimate of the conditional probability that a husband i is nasty to his wife, say, reverting back to our earlier example, given that husband's income is x_i. This is equivalent to the probability that a standard normal variable will be less than or equal to $\alpha+\beta x_i$.

It is possible to show how the probabilities change with the exogenous variables x_i. This can be expressed in terms of rates of change (the derivative of P with respect to the independent variables such as income) as

$$\partial P \ / \ \partial x_j = \beta_j f(Z) \ ,$$

where $f(Z)$ is the value of the normal density function at the point Z. The change in probabilities becomes progressively smaller as one approaches either $P = 0$ or $P = 1$, as one can see from Table 13.1.

Although the probit model is more appealing than the linear probability model, it generally involves nonlinear estimation techniques (because the cumulative normal transformation is nonlinear) and thus excessive computing time and costs. Additionally, given the logit transformation, the theoretical justification for employing the probit model is rather limited in scope.

13.2.4 Logit Transformation

Perhaps the most frequently assumed form for the underlying distribution function is the logistic distribution. The *logit model* is based on the cumulative logistic probability function and is expressed as

$$P_i \equiv \text{Prob}(y_i = 1) = F(z_i) = F(\alpha + \beta x_i) = \frac{1}{1+e^{-z_i}} = \frac{1}{1+e^{-(\alpha+\beta x_i)}} \ .$$

Table 13.2. Values of cumulative probability functions
(each distribution is assumed to have zero mean and unit variance)

z	Cumulative normal $p(Z) = \dfrac{1}{(2\pi)^{1/2}} \displaystyle\int_{-\infty}^{Z} e^{-u^2/2}\, dv$	Cumulative logistic $p'(Z) = \dfrac{1}{1+e^{-Z}}$
-3.0	.0013	.0474
-2.0	.0228	.1192
-1.5	.0668	.1824
-1.0	.1587	.2689
$-.5$	.3085	.3775
0	.5000	.5000
$.5$	.6915	.6225
1.0	.8413	.7311
1.5	.9332	.8176
2.0	.9772	.8808
3.0	.9987	.9526

Here, e represents the base of the natural logarithm and is approximately equal to 2.718. P_i, as before, is the probability that an individual makes a certain choice, given knowledge of x_i. The distribution ranges from 0 to 1 as z_i goes from $-\infty$ to $+\infty$. To get a feel for the cumulative logistic function, it might be useful to examine Table 13.2 and Figure 13.1. Table 13.2 lists values of both the cumulative normal function and the cumulative logistic function for various values of z. As both the table and the graph show, the logit and probit formulations are quite similar; the only difference is that the logistic has slightly fatter tails. In other words, the normal curve approaches the axis more quickly than the logistic curve does. In fact, as Hanushek and Jackson (1977) point out, the logistic distribution closely resembles the student's t-distribution with seven degrees of freedom, while (as is widely known) the t-distribution approximates the normal as the number of degrees of freedom gets sufficiently large. Because it is quite similar in form to the cumulative normal but easier to apply from a computational point of view, the logit or cumulative logistic model is often used in preference to the probit model.

The popularity of the cumulative logistic function arises from the convenient mathematical properties along with its desirable shape. Let us note that if $P_i = \dfrac{1}{1+e^{-z_i}}$, we have

$$1 - P_i = 1 - \frac{1}{1+e^{-z_i}} = \frac{e^{-z_i}}{1+e^{-z_i}} = e^{-z_i} P_i \ .$$

Thus, $e^{-z_i} = \dfrac{1-P_i}{P_i}$. But, since by definition, $e^{-z_i} = 1 / e^{z_i}$, we have $e^{z_i} = P_i / (1-P_i)$. Now, taking natural logarithms of both sides, we have

$$\log e^{z_i} = \log P_i / (1 - P_i)$$

or

$$z_i = \log P_i / (1 - P_i) \ ,$$

where $\log (\cdot)$ denotes the natural logarithm of an algebraic expression, $(\cdot)$. (Note that this is often written as $\log_e(\cdot)$ or $ln(\cdot)$ as opposed to $\log_{10}(\cdot)$.) Thus, we get

$$L \equiv \log P_i / (1 - P_i) = z_i \equiv \alpha + \beta x_i \ ,$$

where *L is called the logit or the log of the odds ratio, and analysis based on this distributional assumption is often termed "logit analysis."*
     The dependent variable in the above equation is thus the logarithm of the odds that a particular behavior will be observed. An important advantage of the logit model is that it manages to transform the problem of predicting probabilities in the [0,1] interval to the problem of predicting the odds of an event's occurring within the entire range from $-\infty$ to $+\infty$; thus, as P goes from 0 to 1 (z goes from $-\infty$ to $+\infty$), L goes from $-\infty$ to $+\infty$; hence, while the probabilities are bounded, the logit or log-odds are unbounded with respect to the values of x.
     The slope of the cumulative logistic function is maximum at $P = 1/2$. In terms of the regression model, this leads to the implication that changes in the independent variable(s) will have their greatest impact on the probability of choosing a given option at around the midpoint of the distribution. The rather narrow slopes near the endpoints of the distribution similarly imply that large changes in x are needed to bring about a small change in probability. This is really because,

although the logits are linear functions of the explanatory variables, the probabilities themselves are not. (As a matter of fact, transformations of the explanatory variables are permitted so that one may regard the logits as linear in the parameters and not necessarily the original variables; in other words, one may replace the x by $g(x)$, where g is one of many feasible transformations like the exponential, reciprocal, or logarithmic, etc.)

The relationship between the change in probabilities and a change in any one of the independent variables can be seen by taking the first-order partial derivative of P with respect to one of the explanatory variables, say x_k, as follows:

$$\frac{\partial P}{\partial x_k} = \frac{\partial (1/(1+e^{-z}))}{\partial x_k} = \frac{\partial}{\partial x_k}\left[\frac{1}{1+e^{-(\alpha+\beta x)}}\right]$$

or,

$$\frac{\partial P}{\partial x_k} = \frac{1}{[1+e^{-(\alpha+\beta x)}]^2}\,(\beta_k)\,e^{-(\alpha+\beta x)}$$

because

$$\frac{\partial(\frac{1}{y})}{\partial x} = -\frac{1}{y^2}\cdot\frac{\partial y}{\partial x}$$

and

$$\frac{\partial e^y}{\partial x} = e^y\,\frac{\partial y}{\partial x}\ .$$

Or,

$$\frac{\partial P}{\partial x_k} = \beta_k\,P(1-P)\ ,$$

because

$$P = \frac{1}{1+e^{-(\alpha+\beta x)}}$$

and

$$1 - P = \frac{e^{-(\alpha+\beta x)}}{1+e^{-(\alpha+\beta x)}} .$$

One can also see that *this form would automatically allow for interactions among the independent variables* because the value of the derivative depends on where it is evaluated (in terms of P) and this level of P will depend on the values of both x_k and any other explanatory variable(s).[3]

Another point worth noting here is that a very serious difficulty would occur if one were to attempt to estimate the equation

$$\log \left[\frac{P_i}{1-P_i} \right] = z_i = \alpha + \beta x_i$$

directly. If P_i happens to equal either 0 or 1, the natural logarithm of the odds will be undefined. Thus, the application of standard regression techniques like ordinary least squares (OLS) is clearly inappropriate, where P_i is set to 1 if a given choice is made and 0 otherwise. *The correct estimation procedure for the logit model can best be grasped by distinguishing between studies in which individual differences are the basic units of analysis (namely, the use of micro data) and studies in which the analysis involves use of grouped data.* We will now go through each in turn.

---

[3]The interaction may be evaluated by finding the second derivative as $\partial^2 P / \partial x_k \, \partial x_l = \beta_l \, \beta_k \, (1-P)(1-2P)$, which is a function of β_k, β_l and P.

13.3 Logit Analysis: Grouping Data Across Subjects within Each Cell

Let us first consider an instance in which information about the frequency of an event's occurring (e.g., husband is nasty) in a given subgroup of the population is available but there is no knowledge of the behavior of every individual in that given subgroup. The problem then translates into one of first measuring the actual probability of a given outcome or event for values of the explanatory variables (e.g., income), and then approximating the functional form of the relationship between such probabilities and the values of the x's.

If we have many observations of the actual outcomes for given values of x, we can calculate the relative frequency of the event's occurring and use this as the probability of the occurrence. This is, in fact, the standard approach to dealing with probabilities in that the probability of an event is the limit (as the sample size gets infinitely large) of the relative frequency of occurrence. This essentially provides the basis for an experimental design approach to grouped logit data.

Even with a nonexperimental setting, where many observations for each value of x are available, we can use a similar approach to estimate the underlying probabilities. For instance, if the explanatory variables are themselves categorical – such as age grouping, race, occupation (blue-collar vs. white-collar), or social class, and so forth – it is possible to observe the frequency of the outcome within each categorical grouping. As the sample size of each subgroup increases, the observed frequency becomes a better estimate of the true probability. When the explanatory variables are not categorical (e.g., actual reported income), we can group them into categories and proceed as before.

Thus, in particular, if for each category x_j $(j = 1, \ldots, J)$ one has n_j observations of a dichotomous (0 or 1) dependent variable $y_{ij}(i = 1, \ldots, n_j)$, the estimated probability for the jth group will be given by

$$\hat{p}_j = \frac{1}{n_j} \sum_{i=1}^{n_j} y_{ij} = \frac{r_j}{n_j}$$

where r_j denotes the total number of times the first alternative ($y = 1$) is chosen by individuals in the group j.

Then it seems reasonable to estimate the logit model (assuming a logistic distribution of the probabilities) by using an estimate of the probability of a given choice *for each group* by identical individuals. The approximation

$$p_j \cong \hat{p}_j = \frac{r_j}{n_j}$$

makes sense if it is taken into account that r_j obeys the binomial probability distribution, for which the mean frequency of occurrence is r_j / n_j.

We can then go on to estimate the logit probability model by using $\hat{p}_j$ to approximate p_j so that

$$\log \frac{p_j}{1-p_j} \cong \log \frac{\hat{p}_j}{1-\hat{p}_j} = \log \frac{r_j/n_j}{1-r_j/n_j} = \log \frac{r_j}{n_j-r_j} = \alpha^* + \beta^* x_j + e_j$$

The equation is linear in the parameters and can be estimated using ordinary least squares regression methodology. For small samples the estimates may be biased, but the results improve remarkably as the number of observations associated with each of the levels of x increases in magnitude. In fact, the estimates are consistent when the sample size *in each group* gets sufficiently large. This requirement for consistency is more stringent than the usual one that the total sample size be large, but is called for to assure that the distribution of observations associated with each group approaches normality.

This method was first proposed by Berkson (1953) and later extended by Theil (1970). Note that this grouping procedure can also be used when individual observations are available. It is most appropriate, however, for situations where there are many observations and where the explanatory (or independent) variables can themselves be classed into natural groupings. Nevertheless, because ordinary least squares is a substantially simpler estimation procedure than the more generalized logit estimation procedure to be described in the following sections, it may be decided to divide the independent variable(s) arbitrarily into groups and to calculate frequencies within each group. Once again, we stress the importance of having a reasonable number of observations within each group so that the estimated frequency provides a good estimate of the true probability.

To illustrate how this procedure might actually apply, let us take a case in which we analyze the husband's behavior toward his wife based on information about his party identification (Democrat, Republican, Independent) and his educational attainment (no college degree, college degree). The data may then be represented as follows:

Party identification

		Democrat	Republican	Independent
Education level	no college degree	$\hat{p}_1$	$\hat{p}_3$	$\hat{p}_5$
	college degree	$\hat{p}_2$	$\hat{p}_4$	$\hat{p}_6$

The $\hat{p}$'s denote the fraction of husbands who are nasty to their wives associated with each of the six possible combinations as represented by the six cells in the above table. For example:

$\hat{p}_1 =$ fraction of Democratic non-college degree holders who are nasty,

$\hat{p}_2 =$ fraction of Democratic college degree holders who are nasty,

.
.
.

$\hat{p}_6 =$ fraction of independent college degree holders who are nasty.

Since there are six categories or groupings defined by the six possible combinations of party identification and education level, the least-squares regression will have six observations. The dependent variable observations will be given by

$$\hat{L}_1 = \log \frac{\hat{p}_1}{1-\hat{p}_1}, \hat{L}_2 = \log \frac{\hat{p}_2}{1-\hat{p}_2}, \ldots, \hat{L}_6 = \log \frac{\hat{p}_6}{1-\hat{p}_6} \ .$$

The independent or explanatory variables in the regression will have to be defined by a series of dummy or indicator variables describing the category to which each observation belongs. Thus, if we let

$$x_2 = \begin{cases} 1 & \text{\textit{for Republican}} \\ 0 & \text{\textit{otherwise}} \end{cases}$$

$$x_3 = \begin{cases} 1 & \text{\textit{for Independent}} \\ 0 & \text{\textit{otherwise}} \end{cases}$$

and

$$x_4 = \begin{cases} 1 & \text{\textit{for college degree holders}} \\ 0 & \text{\textit{otherwise}} \end{cases}$$

the logit model will be *estimated* as

$$\log \frac{\hat{p}_j}{1-\hat{p}_j} = \beta_1 + \beta_2 x_2 + \beta_3 x_3 + \beta_4 x_4 + e_j \ ,$$

where $\hat{p}_j$ is the fraction of nasty husbands in each combination of party identification (Democrat, Republican, Independent) and educational level (college degree, no college degree) in the sample.

It should be pointed out that we have included a constant term but omitted one dummy for each category to prevent singularity or perfect collinearity among the variables. Also, the error term arises, of course, because $\hat{p}_j$ is only an estimate of the true probability p_j.

Assuming for a moment that each $\hat{p}_j$ accurately measures the true frequency in the population of husbands, the interpretation of this logit model with grouped data is quite straightforward. Thus, we have

$\hat{L}_1 \cong L_1 = \beta_1 \equiv$ predicted odds of being a nasty husband by Democratic college non-degree holders,

$\hat{L}_2 \cong L_2 = \beta_1 + \beta_4 \equiv$ predicted odds of being a nasty husband by Democratic college degree holders,

$\hat{L}_3 \cong L_3 = \beta_1 + \beta_2 \equiv$ predicted odds of being a nasty husband by Republican college non-degree holders,

$\hat{L}_4 \cong L_4 = \beta_1 + \beta_2 + \beta_4 \equiv$ predicted odds of being a nasty husband by Republican college degree holders,

$\hat{L}_5 \cong L_5 = \beta_1 + \beta_3 \equiv$ predicted odds of being a nasty husband by Independent college non-degree holders,

$\hat{L}_6 \cong L_6 = \beta_1 + \beta_3 + \beta_4 \equiv$ predicted odds of being a nasty husband by Independent college degree holders.

Thus, if one wants to examine the impact on the husband's nastiness toward his wife of having a college degree, the effect is measured by the value of the coefficient β_4. This impact is the same for any party identification. In other words, the expected difference in the logits between observation on college degree versus no college degree is the same regardless of how the individual is classed on the other explanatory variable, namely, party identification. Such a model is referred to as an "additive model" in classical analysis of variance terminology. Likewise, β_2 measures the difference in the logarithm of the odds (or log-odds) of being nasty between Republican and Democratic husbands. The expected difference in the log-odds between Republican and Independent nasty husbands is $\beta_3 - \beta_2$.

However, since $\hat{p}_j$ does not equal p_j identically, there are certain problems with the use of ordinary least squares estimation in such a grouped-data case. If one assumes that each of the individual observations in a group (or cell) is drawn independently of all others, the observed frequencies $\hat{p}_j$ in each cell follow a binomial probability distribution around the true probability p_j. Then $E(\hat{p}_j) = p_j$, and $var(\hat{p}_j) = p_j(1 - p_j) / n_j$. It then follows directly from the binomial distribution that the estimated dependent variable, $\hat{L}_j = \log \dfrac{r_j}{n_j - r_j}$, will be (asymptotically, i.e., for large samples) approximately normally distributed with mean 0 and variance

$$v_j = \frac{n_j}{r_j(n_j - r_j)} = \frac{1}{\dfrac{r_j}{n_j}(n_j - r_j)} = \frac{1}{\dfrac{r_j}{n_j}\left(1 - \dfrac{r_j}{n_j}\right)}$$

$$= \frac{1}{n_j\hat{p}_j(1 - \hat{p}_j)} \text{, because } \hat{p}_j = \frac{r_j}{n_j} \text{ .}$$

Thus, the error term implicit in the model

$$\log \frac{\hat{p}_j}{1-\hat{p}_j} = \beta_1 + \beta_2 x_2 + \beta_3 x_3 + \beta_4 x_4 + e_j$$

pertinent to the above example is heteroskedastic. In simpler terms, the variance of the error terms for each observation (relevant to each cell or subgroup, namely, 3×2 or 6 in our current example) depends on the probability of occurrence in each cell, $\hat{p}_j$, and on the number of nasty husbands in each cell, r_j. Specifically, the variance is inversely related to the number of observations in each cell, n_j. An obvious correction for such a violation of the OLS regression model is to use a "weighted least squares" regression, where each observation is multiplied by the weight $\frac{1}{\sqrt{v_j}} = \sqrt{n_j \hat{p}_j (1-\hat{p}_j)}$. Another possible correction/adjustment that has been proposed, mainly to help with the small-sample properties of the estimation process, deserves mention here. Suggested by Cox (1970) and also by Domencich and McFadden (1975), it closely resembles the situation of "structural zeros" (and/or small sample problems) in the context of the classical log-linear model. It proposes an improvement in the approximation involved in the model specification by using

$$\log \frac{\hat{p}_j}{1-\hat{p}_j} \cong \log \frac{r_j + 0.5}{(n_j - r_j) + 0.5} \equiv \alpha^* + \beta^* x_j + e_j$$

in place of

$$\log \frac{r_j}{n_j - r_j} .$$

An additional correction for heteroskedasticity has been suggested in which one uses as weights the following as estimates of the error variance:

$$v_j^* = \frac{(n_j + 1)(n_j + 2)}{(n_j(r_j + 1)(n_j - r_j + 1)} .$$

Both adjustments help with small sample estimation but have no impact on the large sample properties of the estimators.

Another alternative way to estimate the model and simultaneously overcome the problem of heteroskedastic error terms is to employ the "generalized least squares" (GLS) estimation technique.

Let us now illustrate how to do hypothesis testing using such a form of logit analysis. If we wish to measure the goodness-of-fit associated with the grouped regression model, we can use an R^2 (coefficient of determination) statistic. However, contrary to the classical regression model, our interest lies not in testing the effect of a single explanatory variable (generally reflected in the significance of the t-statistic relevant to testing for $H_0: \beta = 0$), but rather about alternative specifications of the model. One does this by looking for differences between the actual frequencies in each cell or subgroup and the estimated frequencies from the regression logit model. Specifically, a simple hypothesis we would like to test is whether the difference between the predicted and observed frequencies could occur *by chance* if the estimated model was the correctly specified model. The tests of different alternative specifications are that one given explanatory variable, or set of variables, should be excluded from the model. (We may, of course, test for the significance of interaction terms involving the explanatory variables, deviating slightly from the additive logistic model. This will be covered later in the section.) The null hypothesis is that the effect of these variables or terms is zero. Relevant statistical tests are predicated on the degree of fit of the observed frequencies for each cell to the probabilities predicted by the logit model. Generally, we may infer that the better the fit, the more likely we are to have a correct model.

The development of the appropriate procedures for hypothesis testing will be only briefly outlined here. The statistical justification for the tests depends even more on n_j being large than does the derivation of the estimators for the β's. The essence of the methodology entails that, if each n_j is sufficiently large, the observed frequencies $\hat{p}_j$ are normally distributed about the true probabilities p_j, and that an expression involving their squared difference has a chi-squared distribution. This is because, as n gets large, the binomial distribution approaches the normal. The actual calculations substitute the estimated probabilities from the logit model for the true values; thus, if p_j^* denotes the estimated probability calculated for each observation from the basic logistic model

$$p_j = \frac{1}{1+e^{-z_j}} = \frac{1}{1+e^{-(\alpha+\beta x_j)}} \; ,$$

then the expression

$$\sum_{j=1}^{g} \frac{n_j(\hat{p}_j - p_j^*)^2}{p_j^*(1-p_j^*)}$$

is distributed (asymptotically, i.e., for large samples) according to the chi-squared distribution with degrees of freedom equal to the total number of cells or subgroups/subcategories, g, minus the total number of estimated parameters (see Theil, 1970, and/or McFadden, 1974, for detailed derivations).

The testing for the overall specification of the model essentially involves asking: "What is the probability that the observed frequencies could occur *by chance* if the estimated structure is the correct one?" The better the fit of the model, the smaller the deviations $\hat{p}_j - p_j^*$; and, hence, the smaller the value of the test statistic just mentioned, and the more likely one is *not* to reject the null hypothesis that the given model is the correct one. On the other hand, the worse the fit, the larger the deviations, the higher the X^2 value, and the greater the likelihood that one will reject the null hypothesis.

We may also test to see which of two alternative model specifications provides a better fit to the data. This can be inspected using the difference in the values of the test statistic for the two alternative specifications, based on the additive nature of the likelihood ratio X^2 distribution. The difference chi square, as the difference in the test statistics is often called, is really distributed as chi square with degrees of freedom equal to the number of coefficients being tested for.

This property may be utilized to extend the basic additive model to one involving interaction terms among the explanatory variables. In the example being considered, if we want to extend the hypothesis to one where we would expect husbands with a college degree to be more or less inclined to be nasty to their wives depending on whether they are Republicans or Independents as opposed to being Democrats, we must simply use an additional set of dummy variable terms to allow for such interaction modeling. Thus, the model would become

$$\log \frac{\hat{p}_j}{1-\hat{p}_j} = \beta_1 + \beta_2 x_2 + \beta_3 x_3 + \beta_4 x_4 + \beta_5 x_2 x_4 + \beta_6 x_3 x_4 + e_j \ ,$$

where $\hat{p}_j$, x_2, x_3, x_4 are as before. The new multiplicative explanatory variables $x_2 x_4$ and $x_3 x_4$ are such that they equal 1 for the cells corresponding to Republican college degree holders and Independent college degree holders respectively, and zero for all other cells. Of course, with this model specification, the predicted logit (log-odds), L_1, L_2, and L_5, for non-college degree holders are the same as before, namely, $\hat{L}_1 = \beta_1$, $\hat{L}_3 = \beta_1 + \beta_2$, and $\hat{L}_5 = \beta_1 + \beta_3$. However, the predicted logits for college degree holders turn out to be:

$$\hat{L}_2 \cong L_2 = \beta_1 + \beta_4 \quad \text{(as before for Democrats)}$$

$$\hat{L}_4 \cong L_4 = \beta_1 + \beta_2 + \beta_4 + \beta_5, \text{ and}$$

$$\hat{L}_6 \cong L_6 = \beta_1 + \beta_3 + \beta_4 + \beta_6.$$

Thus, the expected difference in predicted logits between a Republican college degree holder and a Democratic college degree holder is given by $\beta_2 + \beta_5$, whereas the expected difference between a Republican and an Independent is $(\beta_3 - \beta_2) + (\beta_6 - \beta_5)$ if both are college degree holders, and so on.

To test between these two alternative model specifications, namely, the additive and the interaction model, we would use the difference X^2 test. If the difference in the values of the test statistics under the two model specifications exceeds the critical value of the X^2 statistic at a given α-level, say 5% (i.e., $\alpha = .05$), for two degrees of freedom, we would reject the null hypothesis that there is no difference in nasty marital behavior of college degree holders irrespective of their party affiliation.

The logit model being considered is analogous to an analysis of variance model (or a corresponding dummy variable regression model) in the sense that all the exogenous explanatory variables are dummy or indicator variables related to the categories of the explanatory factors. This approach is convenient when there is no naturally occurring scale for the x's – for instance, when one is dealing with a variable like the sex or race of an individual. When at times, however, the x's have a natural scale, such as age or number of children, an approach might be to use a partitioning process to create distinctive categories from the continuous variables.

   Another technique might be to introduce specific values of the variable instead of defining a separate dummy variable for each value of a particular variable. Such additional information can actually be used to increase the efficiency, and often the interpretability, of the parameter estimates. For example, assume that household income and the total number of children are two independent variables used to predict husband's marital behavior. In this case there will be five income categories (namely, $0 to $10,000; $10,001 to $20,000; $20,001 to $30,000; $30,001 to $40,000; and over $40,000) and four categories representing the number of children (say, for example, 0, 1 or 2, 3, 4 or over). The analysis of variance approach would then introduce a separate dummy variable for each category for both the x variables. There will thus be 20 groups of individuals (a person with less than $10,000 income and no children; $10,001 to $20,000 income and no children; and so on). The alternative approach would be to define or introduce only two variables, as follows:

Income   taking on the values of $5,000, $15,000,
                          $25,000, $35,000 and $45,000, depending on
                          which specific category the individual falls
                          under, and

Children   taking on the values of 0, 1.5, 3, 4, 5, say, for
                          the four categories.

The second approach has the advantage of having to estimate far fewer coefficients, which in turn may enhance the precision of the estimates. Furthermore, it provides a way for predicting probabilities for sample individuals in somewhat of an extreme category. For example, if we were interested in predicting the probability of spouse abuse and used an explanatory variable like father's education classified as 10-12 years of education, 13-16 years, 17-18 years, 19+ years, we could estimate the probability of spouse abuse for someone whose father had a PhD, for instance.

   Some potential difficulties may limit the uses of the logit models of the kind just considered. The development of the statistical procedures and the justification for the tests of significance depend to a very great extent on having a large number of individual observations within each cell of the contingency table. A useful rule of thumb for the application of the least-squares approximation is that each cell of the table have at least five observations. A more accurate rule would account for the fact that the least-squares approximation is poorest for levels of x in which the frequency of a given choice is close to either 0 or 1. One may observe from the expression for v_j

$$v_j = \frac{n_j}{r_j(n_j - r_j)} = \frac{1}{\frac{r_j}{n_j} \cdot n_j(1 - \frac{r_j}{n_j})}$$

that v_j gets arbitrarily large as (r_j / n_j) approaches either 0 or 1. This large variation in the frequency estimates guarantees inaccurate parameter estimates, making additional observations desirable. The point to note is that the large sample assumption applies to *each* cell in the contingency table, and not to the total sample size. It so happens in many applications that even though there may be a very large number of total observations, many of the individual cell entries become too small by the time several explanatory variables are included in the contingency table. The problem with small cells is one of inefficiency, however, and not one causing bias in the model.

Another related problem arises from the fact that the approximation implicit in the above modeling is not strictly appropriate when any of the explanatory variables are continuous in nature. The continuous variables must be partitioned before this technique can be applied. Unfortunately, this arbitrary partitioning process can introduce bias through measurement error.

The presence of continuous (or "quantitative" as opposed to "qualitative") variables in models with several attributes as explanatory variables suggests it is necessary to estimate a "logit" model in which only *one* choice is associated with each set of independent variables. Fortunately, in this situation, a different sort of estimation procedure, the maximum likelihood (ML) estimation procedure, can be applied. It does not require the data to be grouped and thus allows each individual observation within the sample to have a distinct probability associated with it.

Since it is possible to prove that a unique maximum always exists for the logit model, ML estimation becomes particularly appropriate and appealing. Almost any nonlinear estimation routine can be used to estimate the model and its associated parameters. The only question becomes, which one of such myriad techniques would lead to a fast convergence of the iterative process and be least expensive in computing costs. It has also been proven that the ML estimates are consistent, and the calculation of the appropriate large-sample statistics is not difficult.

13.4 Logit Analysis without Grouping Subjects: Logistic Regression

To get around some of the problems with the procedures mentioned in the previous section, maximum likelihood estimation techniques have been used that preclude the need to group or categorize the explanatory variables. One immediate advantage is that continuous exogenous variables can be employed as is.

The alternative ML estimator uses the same logistic form for the underlying probability, and one deals directly with the probability function. We estimate the parameters of the logit model

$$p_i = \frac{1}{1+e^{-x_i\beta}}$$

for the probability with $y_i = 1$ for observation i with values of explanatory variables denoted by the vector x_i. Again, as previously noted, the individual p_i's are not observed; instead, there is information for each observation on the behavior of interest (e.g., husband is nasty to his wife). Let us say that the measured dependent variable is $y_i = 1$ if the husband is nasty and $y_i = 0$ if the husband is nice. *The ML procedure differs from the earlier grouped logit model by treating each unit as a separate observation, rather than grouping them to get estimates of p.* The logic behind this is to design an expression for the likelihood of observing the pattern of $y_i = 1$ and $y_i = 0$ in a given sample. Our objective then becomes to find parameter estimates for the β's that make it most likely that the pattern of nasty husband behavior in the sample would have occurred.

If all observations are generated independently, as is reasonable to assume in a cross-sectional analysis, the likelihood of obtaining the given sample of data is equal to the product of the probabilities of the individual observation having the observed outcomes. If we assume that the first husband behavior ($y_i = 1$) is observed n_i times and the second n_2 times, such that $n_1 + n_2 = N$, and if we arbitrarily order the data so that the first n_1 observations are associated with $y_i = 1$, then the likelihood function we wish to maximize has the form

$$L = prob\,(y_1, y_2, \ldots, y_{n_1}, \ldots, y_N)$$

Now, by noting that the probability of $y_i = 1$ is really 1 minus the probability that $y_i = 0$, and using the standard mathematical product notation $\prod$ to denote the product of a number of factors, we find that the likelihood function can be written as

$$L \equiv prob\,(y_i, y_2, \ldots, y_{n_1}, \ldots, y_N)$$

$$= prob\,(y_1)\,prob\,(y_2)\,\ldots\,prob\,(y_N)$$

$$= p_1 \ldots p_{n_1}\,(1 - p_{n_1+1})\ldots(1 - p_N)$$

$$= \prod_{i=1}^{n_1} p_i \prod_{i=n_1+1}^{N} (1 - p_i) = \prod_{i=1}^{N} p_i^{y_i}\,(1 - p_i)^{(1-y_i)}\,,$$

since we have $y_i = 1$ for the first n_1 observations and $y_i = 0$ for the last $n_2(= N - n_1)$ observations.

The above equation indicates the likelihood of obtaining the given sample of data based on the unknown probability elements, p_i. Since the p_i and $(1 - p_i)$ are specific functions of x_i and β's, one obvious implication is that the likelihood function can also be written in terms of x_i and β.

Having chosen the criterion of maximizing the above likelihood function (i.e., the choice is one of the appropriate β's for which L is a maximum), one is simply trying to answer the question: What underlying parameters would be "most likely" to have produced the observed data?

The general approach is to maximize the natural logarithm of the likelihood function instead of the likelihood function itself, since it is somewhat less cumbersome and less difficult. This is possible only because the logarithmic transformation is a monotonic transformation (i.e., it preserves the ordering of the variables after transformation), so that the maximum of $\log L$ occurs at the same values of β as that at which the maximum of L occurs. By thus analyzing $\log L$ one transforms the complicated product of terms in the above equation to a simple sum of logs of the arguments, which is much easier to handle mathematically and in a computer program.

Let us first note that

$$1 - p_i = 1 - \frac{1}{1+e^{-x_i\beta}} = \frac{1+e^{-x_i\beta}-1}{1+e^{-x_i\beta}}$$

$$= \frac{e^{-x_i\beta}}{1+e^{-x_i\beta}} = \frac{1}{1+e^{x_i\beta}} \ .$$

Thus the log-likelihood function for the logistic model becomes

$$\log L = \sum_{i=1}^{N} y_i \log p_i + \sum_{i=1}^{N} (1 - y_i) \log (1 - p_i)$$

$$= \sum y_i \log p_i - \sum y_i \log (1 - p_i) + \sum \log (1 - p_i)$$

$$= \sum y_i \log \frac{p_i}{1-p_i} + \sum \log (1-p_i)$$

Now

$$\log \frac{p_i}{1-p_i} = \log p_i - \log (1-p_i)$$

$$= -\log (1+e^{-x_i\beta}) - [\log (e^{-x_i\beta}) - \log (1+e^{-x_i\beta})]$$

$$= -\log e^{-x_i\beta} = x_i\beta \ .$$

Hence,

$$\log L = \sum_{i=1}^{N} y_i x_i \beta - \sum_{i=1}^{N} \log (1 + e^{x_i \beta}) \ .$$

To obtain the estimators for β that maximize $L^* = \log L$, we take the first order partial derivatives of L^* with respect to the vector of β's, and set such terms equal to zero, as follows:

$$\frac{\partial L^*}{\partial \beta_K} = \sum y_i x_{ik} - \sum \frac{x_{ik} e^{x_{ib}}}{1 + e^{x_i b}}$$

$$= \sum y_i x_{ik} - \sum \frac{x_{ik}}{1 + e^{-x_i b}} = 0 \ for \ k=1, \dots, K \ ,$$

if there are K exogenous variables in the model, giving K nonlinear simultaneous equations that are to be solved for estimated values of β.

The maximum likelihood (ML) estimation techniques can be applied where there are truly categorical explanatory variables, such as sex (male or female), or class (white-collar or blue-collar); or where the only data available have been previously categorized, as with the party identification variable (Democrat, Republican, or Independent, for instance) or with a grouping variable based on marital satisfaction (high or low); or where continuous measures (such as age or a measure of social isolation) and categorical variables are mixed together. *When all the exogenous variables are categorical, the analysis essentially becomes consistent with Goodman's work (1970, 1972a) on the maximum likelihood estimation of log-linear analysis of contingency tables.*

The full maximum likelihood estimator just considered solves several of the problems associated with the grouped logit model (solved earlier by "weighted least squares" or WLS). This method does not require prior grouping and aggregation of explanatory variables, thus avoiding the introduction of measurement bias and the assumption that observations with different characteristics (but falling in the same cell of the contingency table) have identical probabilities. It also does not require very large cell sizes and uses the information in small cells. However, justification for the method is asymptotic, and therefore relies on large sample properties. It is generally accepted, however, that ML estimators have a number of very desirable statistical properties even when applied to small samples. All parameter estimators are consistent and asymptotically efficient. They are also known to be asymptotically normal, so that analogs of the regression t-tests for testing significance of individual coefficients can be applied. Thus, the ratio of the estimated coefficient and its estimated standard error is assumed to follow a normal distribution. If one wishes to test the significance of all or a subset of the coefficients in a logit model with ML estimation, a test based on the LRX^2 replaces the usual F-test in regression.

For instance, suppose we wish to test the significance of the entire logit model. First, we evaluate the likelihood function L when all parameters (other than the constant that people typically include) are set equal to zero. If we call this initial value L_0, and call the likelihood function at its maximum L_{max}, then we define the likelihood ratio as

$$\lambda = \frac{L_o}{L_{max}} .$$

The appropriate test then follows directly from the fact that

$$-2 \log \lambda = -2(\log L_o - \log L_{max})$$

follows a chi-squared distribution with k degrees of freedom, where k is the number of parameters in the equation (other than the constant).

One can use the ML results to obtain a measure of goodness-of-fit analogous to the R^2 in ordinary regression models. One possible simple option is to calculate the value for the expression

$$1 - \frac{L_o}{L_{max}} .$$

This statistic will be zero when the unconstrained likelihood function is no greater than the likelihood function when all parameters are constrained to be zero, and will increase to a maximum value of 1 corresponding to L_{max}.

Another option is to calculate residuals as follows:

$$\hat{e}_i = y_i - \hat{p}_i \ .$$

These residuals will be smaller in absolute value the better the logit model explains the choices being made. From these residuals, one can also calculate an analog to the regression-based R^2. We have

$$ESS = \sum_{i=1}^{N} \hat{e}_i^2 \ , \quad and$$

$$TSS = \sum_{i=1}^{N} (y_i - \bar{y})^2 \ ;$$

$$thus, \quad \hat{R}^2 = 1 - \frac{ESS}{TSS}$$

A computer program for stepwise logistic regression is the BMDP program called PLR (see Appendix 13.2).

13.5 Polytomous Variables: Multiple Choice Models

Let us consider a situation where there are more than two possible categories of behavior. For example, affect in a marital interaction situation may be observed as positive, neutral, or negative. This situation, called polychotomous or polytomous (as opposed to the dichotomous situation where affect may be coded as positive or negative), is a natural extension of the processes previously discussed.

The trick of using dummy variables involves the imposition of no scale or unit of measurement on the variable. However, we must specify a metric when we attempt to extend a single variable to measure polytomous choices. For example, with the affect coding situation, we may define $y = 0$ if the affect is negative, $y = 1$ if neutral, and $y = 2$ if positive. This would imply that the difference between positive and neutral is the same as the difference between neutral and negative, and that "positive" is twice "negative." Thus, we can see that although we can get by with defining variables indicating only qualitative differences in the case of dichotomies, we are forced in the case of polytomous situations to use a variable that indicates "quantitative" differences. Often this is not possible, and in such cases the least squares methodology is not applicable. If all possible choices are mutually exclusive, there are several ways to handle this, depending on certain statistical assumptions and on whether or not a natural ranking or ordering can be associated with each possible choice.

By employing the extension of the logit model in the dichotomous situation to a polytomous one, we can handle any categorical dependent variable; however, we cannot readily incorporate information on *ordering* if such information were available. Goodman (1984) and Agresti (1984) have addressed this problem in extending the approach to the analysis of such *ordinal* categorical data. We will focus here only on the case in which the behavioral alternatives are unranked.

In logit models, the estimated functions express the natural logarithm of the odds of one possible outcome versus another, called the "conditional logit," as a linear function of the explanatory variables. With those possible choices, say $y_j = 1, 2,$ or 3 with respective probabilities of p_j, $j = 1, 2, 3$ (i.e., p_1, p_2, p_3), and dropping the subscript i, denoting individual observations (for notational simplicity), we express the conditional logits as

$$L_1 = \log (p_2/p_1)$$

and

$$L_2 = \log (p_3/p_1)$$

where the odds ratios (p_2/p_1) and (p_3/p_1) represent the odds that $y = 2$ rather than $y = 1$ and that $y = 3$ rather than $y = 1$, respectively. Note that the conditional logits are simply natural logarithms of these conditional odds. Thus,

$$L_3 = \log (p_3/p_2) = \log [(p_3/p_1)(p_1/p_2)]$$

$$= \log [(p_3/p_1) / (p_2/p_1)]$$

$$= \log (p_3/p_1) - \log (p_2/p_1)$$

$$= L_2 - L_1 .$$

Thus, it is not necessary to estimate each situation separately, because the choice of the logit formulation forces constraints on the model.

The logistic model expresses the conditional logits as linear functions of the explanatory variables. Thus

$$L_1 = \log (p_2/p_1) = x\beta_1$$

and

$$L_2 = \log (p_3/p_1) = x\beta_2 \ .$$

Since

$$p_1 + p_2 + p_3 = 1$$

and we have

$$p_2 = e^{x\beta_1} p_1 \quad and \quad p_3 = e^{x\beta_2} p_1 \ ,$$

we get

$$1 = p_1 + e^{x\beta_1} p_1 + e^{x\beta_2} p_1 \ ,$$

or

$$p_1 = \frac{1}{1 + e^{x\beta_1} + e^{x\beta_2}} \ .$$

The polytomous logit model can be estimated either by generalized least squares (GLS) or by maximum likelihood (ML) methods. The GLS estimation technique requires that sufficient repetitions are available so that the observations may be organized in the contingency table format, just as with the dichotomous situation limiting one to categorical explanatory variables.

13.6 Incomplete Tables: Problem of Empty Cells

It often happens that one or more cells in a contingency table are empty, that is, they have observed frequencies of zero. Such empty cells occur for two quite distinctively different reasons.

First, certain combinations of categories may be empty because of logical or definitional considerations. Empty cells of this sort are generally called *structural (or logical) zeros*. Examples may be the classical absence of "pregnant males," or the diagonal nonentries for a table of consequent by antecedent affect for a transition matrix constructed from event sequence data (in which similar events cannot logically follow). The diagonal entries are called "structural zeros" because it is assumed here by definition that one code cannot follow itself. Log-linear models have been adapted to fit tables containing structural zeros. See Bishop et al. (1975), Fienberg (1977), and Haberman (1979); we will not discuss this in detail here.

Second, zero frequencies may also arise by chance, when a particular cell occurs rarely in the actual population, or when the sample size is relatively small compared to the number of cells in the table or for a combination of these reasons. Empty cells of this latter type are generally termed *sampling (or random) zeros*, and one way to avoid them is to pick relatively larger samples. However, there are ways to adjust the classical logit/log-linear models to handle such random zeros in the data.

The models we have considered are all appropriate for tables with sampling zeros, and any such model may be fit to these tables without much trouble, as long as *no* marginal, fit under the model, has a zero entry. We may, indeed, regard the expected frequencies for a well-fitting saturated model as a means of smoothing out uneven observations in the observed table owing to sampling variability. If we want to estimate a model in which there exists a zero marginal (a saturated model, for instance), we may proceed to adjust the cell count to make them all positive. Goodman (1970) suggests adding a small positive number, say one-half (.5), to each cell of the observed table. There are also more complex methods to adjust for sampling zeros (see Bishop et al., 1975).

13.7 Example

In our log-linear analysis of the marital satisfaction-by-social-class data, contrasts (Upton, 1978, p. 63) applied to the RS subtable (consequent affect by marital satisfaction) show that unhappily married husbands display more negative affect than happily married husbands ($z = 4.02$, $p < .0001$), and a similar result holds for wives ($z = 3.79$, $p < .001$). This replicates a well-known result (see Billings, 1979; Gottman, 1979a; Margolin & Wampold, 1981; Schaap, 1982).

The RC table shows that blue-collar husbands display more negative affect than white-collar husbands ($z = 4.10$, $p < .0001$), whereas there is no significant class difference in negative affect for wives ($z = 0.48$, n.s.). To understand this difference, we employed our job distress questionnaire: 2×2 ANOVAs for wife job distress showed no collar effect ($F(1, 47) = 0.36$, n.s.), whereas for husbands there was a significant collar effect ($F(1, 47) = 7.06$, $p < 0.05$). The mean job distress measure for blue-collar husbands was 66.01, and the white-collar mean was 56.35. Although still speculative, this provides some support for the hypothesis that higher job distress is related to husband negative affect for blue collar males.

To continue assessing the relationship between social class, husband negative affect and the husband job distress variable, two analyses were performed: (1) a 2×2 ANOVA with the proportion of husband negative affect as the dependent variable (the design was social class by marital satisfaction); and, (2) a 2×2 ANCOVA with the husband job distress variable as a covariate. There was a significant main effect for social class ($F(1, 48) = 4.38$, $p < .05$), which became nonsignificant ($F(1, 47) = 2.68$, n.s.) when the husband job stress variable was first entered as a covariate. This provides support for the contention that job stress differences between blue- and white-collar men are related to their noncontingent negative affect (unconditional probability).

Our hypothesis was that the couple's philosophy of marriage (which we called "communication orientation") would be an important variable in understanding the negative affect reciprocity effect. Recall that people who score high on this variable report are likely to engage in discussion of disagreements ("engagers"), whereas people who score low on this variable report are likely to avoid these discussions ("avoiders"). We used a median split within each level of marital satisfaction, collapsing across social class, separately for both the husband and the wife. To assess the effect of this variable, we first estimated its effect on homogeneity and found that the ratios of LRX^2 to their df's fell from approximately 2.0 to approximately 1.7. Next, we conducted a multinomial logit-linear fit using maximum likelihood estimation (SAS Institute, 1985), predicting the consequent affect (R) after statistically partialing out the effects of antecedent affect (P), marital satisfaction (S), social

Table 13.3. Effects of the couple's philosophy
of marriage on post affect (R)

Source	df	lRX^2	Probability
Intercept	5	0.2	.999
Preaffect (P)	25	1292.8	.000
Marital satisfaction (S)	5	0.0	1.000
Social class (C)	5	0.0	1.000
P x S interaction	25	49.0	.003
P x C interaction	25	34.7	.093
S x C interaction	5	36.1	.000
Husband philosophy	5	51.8	.000
Wife philosophy	5	35.1	.000

class (C), and the PS, PC, and SC terms. Table 13.3 presents a summary
of this analysis. The effects of both the husband's and the wife's philosophy of marriage were statistically significant.

Appendix 13.1
The Relationship Between Log-Linear and Logit Models

When all variables are categorical or discrete (qualitative) in nature, their
joint sample distribution defines a *contingency table* or a *cross-
classification*, where each combination of variable categories is observed
more than once. Log-linear models are used to model the association
among variables in a contingency table. However, they do not distinguish between dependent and independent variables, in the sense that
there really is not one dependent variable of interest, although there is a
very convenient relationship between log-linear and logit models, as we
shall now see. Indeed, since we most often deal with applications where
one variable is treated as the dependent variable of interest (e.g., consequent affect in marital interaction), we generally employ log-linear
models as a convenient means of fitting equivalent logit models where
the independent variables are categorical. When one of the variables in a
contingency table is deemed to be the dependent variable, the log-linear
model for the table implies a logit model, the parameters of which bear a
very simple relation to the parameters of the log-linear model.

This allows one to use log-linear models as a convenient, although indirect, means of fitting logit models when all the independent variables are qualitative. We shall illustrate the case for a dichotomous dependent variable and later extend the generalization of our findings to the polytomous situation.

Let us consider a saturated log-linear model for a three-way table as denoted by

$$\lambda_{ijk} = u + \alpha_{1(i)} + \alpha_{2(j)} + \alpha_{3(k)} + \alpha_{12(ij)} + \alpha_{13(ik)} + \alpha_{23(jk)} + \alpha_{123(ijk)} \ .$$

If we assume, for the purpose of our example, that the dependent variable is variable number 3, and let Ω_{ij} denote the dependent variable logit with categories i, j of the independent variables, and let π denote probability, we have

$$\Omega_{ij} = \log \frac{\pi_{ij1}}{\pi_{ij2}} = \log \frac{n\pi_{ij1}}{n\pi_{ij2}}$$

$$= \log \frac{u_{ij1}}{u_{ij2}} = \lambda_{ij1} - \lambda_{ij2} \ .$$

Thus, for the standardized log-linear model under consideration, we have

$$\Omega_{ij} = (\alpha_{3(1)} - \alpha_{3(2)}) + (\alpha_{13(i1)} + \alpha_{13(i2)})$$

$$+ (\alpha_{23(j1)} + \alpha_{23(j2)}) + (\alpha_{123(ij1)} - \alpha_{123(ij2)}) \ .$$

We note that $(\alpha_{3(1)} - \alpha_{3(2)})$ does not depend on the independent variables, the term $(\alpha_{13(i1)} - \alpha_{13(i2)})$ depends only on variable number 1, and so on. We can rewrite the equation as

$$\Omega_{ij} = \omega + \omega_{1(i)} + \omega_{2(j)} + \omega_{12(ij)} \ ,$$

where we have essentially (a set of ANOVA-like constraints on the α's):

Table 13.4. Comparison of logit models and corresponding
log-linear models for a 3-way table

Logit model	Log-linear model	Marginal fit
$\omega_{ij} = w$	$\lambda_{ijk} = u + \alpha_{1(i)} + \alpha_{2(j)} + \alpha_{3(k)} + \alpha_{12(ij)}$	[12] [3]
$\omega_{ij} = w + w_{1(i)}$	$\lambda_{ijk} = u + \alpha_{1(i)} + \alpha_{2(j)} + \alpha_{3(k)} + \alpha_{12(ij)} + \alpha_{13(ik)}$	[12] [13]
$\omega_{ij} = w + w_{2(j)}$	$\lambda_{ijk} = u + \alpha_{1(i)} + \alpha_{2(j)} + \alpha_{3(k)} + \alpha_{12(ij)} + \alpha_{23(jk)}$	[12] [23]
$\omega_{ij} = w + w_{1(i)} + w_{2(j)}$	$\lambda_{ijk} = u + \alpha_{1(i)} + \alpha_{2(j)} + \alpha_{3(k)} + \alpha_{12(ij)} + \alpha_{13(ik)} + \alpha_{23(jk)}$	[12] [13] [23]
$\omega_{ij} = w + w_{1(i)} + w_{2(j)} + w_{12(ij)}$	$\lambda_{ijk} = u + \alpha_{1(i)} + \alpha_{2(j)} + \alpha_{3(k)} + \alpha_{12(ij)} + \alpha_{13(ik)} + \alpha_{23(jk)} + \alpha_{123(ijk)}$	[123]

$$\omega \equiv \alpha_{3(1)} - \alpha_{3(2)} = 2\alpha_{3(1)}$$

$$\omega_{1(i)} \equiv \alpha_{13(i1)} - \alpha_{13(i2)} = 2\alpha_{13(i1)}$$

$$\omega_{2(j)} \equiv \alpha_{23(j1)} - \alpha_{23(j2)} = 2\alpha_{23(j1)}$$

$$\omega_{12(ij)} \equiv \alpha_{123(ij1)} - \alpha_{123(ij2)} = 2\alpha_{123(ij1)} \; .$$

Moreover, because the ω's are defined as twice the α's, they are also constrained to sum to zero over any coordinate, so that

$$\omega_{1+} = \omega_{2+} = \omega_{12(i+)} = \omega_{12(+j)} = 0 \; \textit{for all i and j} \; .$$

Let us consider Table 13.4, summarizing logit models, the corresponding log-linear models, and the fitted margins for the three-way table under consideration.

We see that the log-linear model parameters for the association of the independent variables do not appear in the equations for the corresponding logit models. The dependent-variable log-odds thus depend on the independent variables and their interactions for such a saturated model.

Similarly, the same argument applies with respect to any unsaturated log-linear model, each of which implies a model for the dependent variable logits. If our purpose is to examine the effect of the independent variables on the dependent variable, and not to explore the relations among the independent variables, we would generally include the α_{12} in this case of a three-way table and its lower-order relatives, namely, α_1 and α_2, in any model we fit, thereby treating the associations between the independent variables as given. Indeed, the independent-variable marginal table is often fixed by design in a typical experimental situation. Fienberg (1977) notes that when data are sparse, it may be better to smooth the relationships among the independent variables by not treating their marginal table as fixed. However, we will generally consider the independent-variable marginal to be fixed.

Sample logits are very analogous to cell means in ANOVA, and thus deserve close examination. It does not make much sense to report significance tests for the parameters of a logit model in the absence of cell logits, or the parameter estimates themselves, just as it is not too enlightening to report an analysis of variance table without cell means. It is good, therefore, to plot the sample logits against the levels of the independent variables (factors), just as we do in the ANOVA situation.

The relationship between log-linear and logit models may be properly extended to polytomous dependent variables. Let us consider, for instance, a three-way table of consequent affect (+, 0, or −) [variable 3], antecedent affect (+, 0, or −) [variable 2] and the degree of marital satisfaction (high or low) [variable 1]. Treating consequent affect as the dependent variable, we may restrict our attention to only models that fit the {12} marginal [antecedent affect by marital satisfaction level]. These log-linear models then become identical to logit models for the three-category dependent variable polytomy.

It is alternatively feasible to fit log-linear/logit models for nested dichotomies. If we were considering, for instance, a model for predicting the respondent's party identification (Democrat, Republican, or Independent) based on the respondent's parents' party identification (both Democrats, both Republicans, other) and on parents' level of political activity (both active, one active, and neither active), this approach would seem preferable. It appears natural to test for the following dichotomies: {2, 13} Independents versus party identifiers; and {1, 3} Democrat versus Republican identifiers.

Appendix 13.2

Computer Programs for Data Analysis

The three widely used strategies for fitting log-linear and logit models are:

1. Iterative proportional fitting (IPF) of hierarchical log-linear models, analogous to hierarchical analysis of variance models. Marginal tables sufficient for model parameter estimation are used to obtain maximum likelihood (or minimum-discrimination information estimates and/or Pearson chi-squared test statistics; see Ku & Kullback, 1974, and Ku & Kullback, 1968. This method involves applying the algorithm of Deming and Stephan (1940). This algorithm was shown to be relevant for log-linear model fitting by Darroch (1962), Birch (1963), Bishop (1969); it was described by Goodman (1970), and by Bishop et al. (1975).

2. Weighted least squares (WLS) fitting of asymptotic regression models to log-linear functions of observed cell frequencies, which lead to the linearized minimum modified Neyman X^2 estimation and Wald statistics. This method was actually developed by Grizzle and Williams (1972) as a specific application of the more generalized procedure of Grizzle, Starmer and Koch (1969) (usually referred to as GSK in the literature). It is based on earlier works of Wald (1943), Neyman (1949), Bhapkar (1961, 1966, 1970) and the literature on the logistic models due to Berkson (1944, 1953). WLS was also used by Theil (1970) within the context of qualitative dependent variables in the field of econometrics.

3. Maximum likelihood (ML) techniques of a more general nature, for example, Newton-Raphson or iteratively reweighted least squares or modifications thereof, which lead to the same type of solutions as those obtained by IPF methods. Among the major writers responsible for the recent popularity of this method are Cox (1970), Gokhale (1972), Nelder and Wedderburn (1972), Haberman (1974), Nelder (1974), and Bock (1975).

Among the readily available computer programs implementing Method #1 are the ECTA program due to Goodman and Fay (1973) utilizing the method of Haberman (1978, 1979), the SPSS[X] procedure HILOGLINEAR (SPSS Inc., 1986), and the BMDP procedure P4F (Dixon, 1985), also implementing the model selection techniques proposed by Brown (1976) and Bendetti and Brown (1978) as an extension to stepwise regression models. BMDP-4F, unlike generally used IPF procedures, is also programmed to utilize the information matrix, and thus provides estimated asymptotic covariance matrices for log-linear model parameters or fitted counts. This method is, however, necessarily limited to the *hierarchical* class of ANOVA type of models due to Bishop et al. (1975).

Examples of software in the Method #2 group include, among others, GENCAT (Landis et al., 1976), NONMET, and the more commonly used SAS procedure FUNCAT (SAS Institute, 1985) which has recently been modified and renamed as CATMOD as part of SAS Version 5 software (SAS Institute, 1985). Here, asymptotic covariance matrices are obtained, allowing one-pass execution of multistage modeling procedures.

Representing the group of Method #3 software are the British package GLIM (Baker & Nelder, 1978), Bock's package called MULTIQUAL (Bock & Yates, 1973), the SPSS[X] procedure LOGLINEAR (SPSS Inc., 1986; Norusis, 1985), and, once more, the SAS procedures FUNCAT and CATMOD as outlined under Method #2 (SAS Institute, 1985). These methods also use multistage modeling procedures.

It should be noted here that both methods #2 and #3 allow hierarchies of models to be fitted without repeated reconstruction of the full design matrix. Also, within the context of a hierarchical ANOVA class of log-linear models, methods #1, #2 and #3 are statistically equivalent in large sample situations, in the sense that analogous vectors of estimators and test statistics from these respective methods have the same asymptotic distributions for purposes of inference.

14

THE PROBLEM OF AUTOCONTINGENCY

AND ITS SOLUTIONS

14.1 The Problem

This chapter outlines a few approaches to the basic problem of autocontingency. What is the problem? Briefly, in sequential analysis we want to determine if we can predict the occurrence of code B at time $t + 1$ by knowing that code A occurred at time t. Usually we have a dyad we are observing, and the B codes come from observing one person's behavior while the A codes come from observing the other person's behavior. The standard we employ to evaluate the prediction is the amount of improvement over the unconditional probability of B, $p(B)$. However, there is another standard that makes a lot more sense when we try to establish a causal link between A and B. That standard follows from the fact that knowing that B occurred at time t would lead, in many cases, to gain in prediction over and above the unconditional, $p(B)$. Hence, the test for sequential connection could be made more stringent by comparing $p(B \mid A)$ not with $p(B)$ but with the residual from $p(B)$ obtained from predicting B's occurrence by B's own past behavior. This could be viewed simply as an adjustment to $p(B)$ in the numerator of our usual z-scores for sequential connection. The nature of this adjustment is the subject of this chapter. Currently, we have little experience with such solutions, and thus our review is entirely theoretical. As we gain experience, we will be able to say more about the relative merits of the alternative reviewed here. We now turn to the nature of the problem of autocontingency.

Suppose there are two time series, one for the mother's behavior and one for the infant's behavior (see Figure 14.1). We want to make an inference about the influence of mother on baby and baby on mother. This is similar to the problems political scientists have in inferring

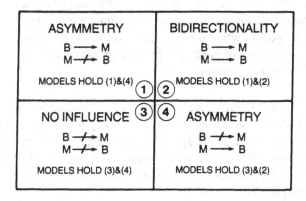

Figure 14.1. Four quadrants illustrating bidirectionality, dominance (or asymmetry in predictability), and no interaction for two individuals, M (mother) and B (infant) (from Gottman & Ringland, 1981).

relationships between two time series, such as military spending in the USSR and that in the United States. Gottman and Ringland (1981) wrote:

What needs to be demonstrated is more than that predictability (at any lag) exists from mother to infant and from infant to mother. By itself this demonstration is inadequate to infer bidirectionality because of the problem of auto-correlation. This was recently pointed out by Sackett (1980), who wrote, "The basic issue of autocontingency has not been addressed by students of social interaction. Unfortunately, autocontingency does affect the degree to which crosslag dependencies can occur. In some instances, apparent cross-contingencies may be a total artifact of strong autolag functions" (p. 330). Sackett's point is consistent with Jenkins and Watts (1968), who showed that "very large cross covariances, all of them spurious, can be generated between two uncorrelated processes as a result of the large autocovariances within the two processes" (p. 338). To demonstrate that the behavior of the baby is influenced by the behavior of the mother, this paper suggests that we need to show that we can reduce uncertainty in the infant's behavior, over and above our ability to predict simply from the infant's past. A similar discussion can be found in a recent paper by Pierce (1977) on assessing causal relationships between economic time series. Bidirectionality occurs when we can demonstrate the converse as well, and asymmetry in predictability occurs when we cannot demonstrate this symmetry.

We must *control for* autocorrelation to infer cross-correlation. In the military spending example we know that each country's military spending is affected to some extent by its spending last year and by common factors such as inflation. That is not what we are interested in. We want to assess the extent of influence between the two countries over and above the influence of their own past.

14.2 Sackett's Computational Solution

Sackett (1980) analyzed talk or silence data from 16 researchers at a conference in Madison, Wisconsin, in which they discussed social interaction research. Sackett computed autolags for 16 speakers at the conference (minus Sackett, who was busy at the self-appointed task of coding their behavior). Sackett noted that, over all autolags, speaker N followed himself with an average probability of 29.8%. However, N spoke only 2.3% of the 1,506 speakerships (or 35 times). Nonetheless, once N has spoken, the chance of others to follow is much lower than expected from their unconditional probabilities. Because N followed himself 29.8% of the time, other people could only speak 70.2% of the time. However, the unconditional probabilities of these 15 speakers sum to $100-2.3 = 97.7\%$. Hence, speaker N's autocontingency has restricted the chance for others to speak following him by $97.7-70.2 = 27.5\%$. In the same way, a *negative* speaker autocontingency makes the unconditional probabilities that others will speak an *underestimate*. Sackett wrote:

> In sum high autolag magnitude and cyclicity make simple unconditional probabilities inadequate as expected values for relationships between an autocontingent event and other events.... Excited autolag probabilities will yield too few significant crosslags and inhibited autolag probabilities too many significant crosslags when unconditional probabilities are used as expected values. (p. 319)

The Sackett solution to the autocontingency problem is the following. We want to test the probability that speaker X will follow speaker N (these could be *any* two codes). Ordinarily we would form a z-score comparison of $p(X \mid N)$ and $p(X)$. However, $p(X)$ is not an appropriate comparison, so it must be adjusted for N's autocontingency.

Bakeman (1987, personal communication) suggested that the adjustment can be made by running his program ELAG4 in two ways: once in which codes can follow themselves and once in which they cannot. For example, if we have the following marital interaction data (1 = HUSBAND POSITIVE, 2 = HUSBAND NEUTRAL, 3 = HUSBAND NEGATIVE, 4 = WIFE POSITIVE, 5 = WIFE NEUTRAL, 6 = WIFE NEGATIVE):

```
2 2 4 4 5 1 5 3 6 2 5 2 6 2 3
6 2 5 2 6 2 6 2 5 2 5 2 5 5 2
2 6 2 5 3 2 5 2 5 3 2 5 3 2 5
6 2 6 2 5 1 6 2 3 2 6 3 3 5 2
5 2 6 2 5 5 5 5 5 5 2 3 2 5
2 6 2 6 2 6 6 4 2 5 5 6 1 5 1
5 6 5 5 6 5 2 5 2 6 2 3 2 2
6 6 2 6 3 2 2 6 2 5 2 5 5 2 5
2 5 2 6 2 3 3 2 2 6 2 6 6 6 5
6 6 6 2 6 2 5 2 5 2 5 2 5 5 6
6 2 4 5 6 2 5 5 2 5 2 5 2 4 5
2 5 2 5 2 5 2 5 2 5 1 6 6 6 6
2 2 4 2 5 2 2 5 5 2 5 2 5 2 6
2 6 6 6 2 5 2 5 2 2 5 2 5 2 5
2 5 5 2 5 5 2 5 3 5 2 3 5 3 2
6 2 5 2 2 5 3 4 2 5 5 2 5 2 5
2 5 2 5 5 6 2 2 4 5 2 5 2 5 2
5 2 2 5 2 5 2 5 5 2 5 2 5 5 2
5 2 5 5 4 4 5 5 5 4 5 5 2 4
```

Suppose we are interested in the conditional probability $p(HNEU \mid WNEG) = 14/43 = .558$. The unadjusted unconditional probability is $p(HNEU) = .372$; the adjusted unconditional is .409. This latter figure is obtained by running ELAG4 with the option that codes cannot follow themselves. If we do not control for autocorrelation, we would obtain ($B = HNEU$, $A = WNEG$):

$$z = \frac{p(B \mid A) - p(B)}{\sqrt{\dfrac{p(B)(1-p(B))(1-p(A))}{(n-1)(p(A))}}}$$

$$= \frac{.558 - .371}{\sqrt{\dfrac{(.372)(1-.372)(1-.160)}{(269-1)(.160)}}}$$

$$= 2.74$$

If we adjust for autocorrelation, we would obtain:

$$z = \frac{.558 - .409}{\sqrt{\dfrac{(.409)(1-.409)(1-.141)}{(220-1)(.141)}}}$$

$$= 1.82$$

14.3 The Logit Linear Solution

Allison and Liker (1982) suggested a logit analysis solution for the problem of autodependence. First they discussed a chi-squared test.

14.3.1 Chi-Squared Test

Suppose we want to estimate the effect of H_t on W_{t+1}, controlling for W_t. For this analysis a full three-way table is needed. The hypothesis can be written as:

$$Pr[W_{t+1} = 1 \mid W_t, H_t] = Pr[W_{t+1} = 1 \mid W_t] \ .$$

This equation says that W_{t+1} may depend on W_t but not on H_t. We can treat the following table, from Allison and Liker, as two 2×2 tables (see Tables 14.1 and 14.2). The test is performed by computing a chi-squared test for independence in each 2×2 subtable and summing these two values.

Table 14.1

		W_{t+1}	
W_t	H_t	1	0
1	1	577	139
1	0	222	76
0	1	169	1089
0	0	149	839

Table 14.2

	$W_t = 1$		$W_t = 0$	
	W_{t+1}		W_{t+1}	
H_t	1	0	1	0
1	577	139	169	1089
0	222	76	149	839

For the data in Tables 14.1 and 14.2, Pearson $X^2 = 5.91$, df $= 2$, not significant at alpha equal to .05. Thus, when the autodependence is partialed out, the cross dependence is greatly reduced.

14.3.2 Logit Linear Model

An alternative is to fit a logit-linear model by the methods of maximum likelihood. The null hypothesis model is:

$$logit\ [Pr(W_{t+1} = 1 \mid W_t, H_t)] = \alpha + \beta\ W_t\ .$$

The advantage of this approach is that it can handle complex forms of autodependence. For example, we can allow for autodependence up to lag 3 by fitting the model:

$$logit \, [Pr(W_{t+1} = 1 \mid H_t, W_t, W_{t-1}, W_{t-2})] =$$

$$\alpha + \beta_1 H_t + \beta_2 W_t + \beta_3 W_{t-1} + \beta_4 W_{t-2} \, .$$

This model would be fit to the five-way table for W_{t+1}, W_t, W_{t-1}, W_{t-2}, and H_t.

14.4 The Gardner-Hartmann Correction

Gottman and Ringland (1981) solved this problem for continuous time-series data. Let us examine the problem for binary data. Gardner, Hartmann, and Mitchell (1982), in a Monte Carlo simulation study, tested whether the Pearson chi-squared statistic to evaluate contingency between two behaviors in a dyad is affected by three factors: (1) the dependence of each person's behavior on the past of that behavior; (2) the base rates of occurrence of the behaviors; and (3) the length of the simulated series. They used the lag-one phi coefficient as an index of autocontingency, and they found that autocontingency strongly affected Type I error rates. When autocontingency increased, Type I error rates also increased. In a later paper, Gardner and Hartmann (1984) reported that these error rates can also be affected conservatively (i.e., decreased) or unaffected, depending on the nature of autocontingency. The role of base rates on Type I error was low at moderate base rates (.25 to .75) but it increased at extreme base rates (.05 to .95). The length of the series had no effect.

In a later paper Gardner and Hartmann (1984) reported a result of Tavare and Altham (1983) that if ϕ_M and ϕ_I index the lag-one autocontingency for the mother and baby, respectively, the X^2 statistic for inferring cross-contingency between mother and infant should be multiplied by the parameter

$$(1 - \phi_M \, \phi_I) / (1 + \phi_M \, \phi_I)$$

The index of autocontingency is the phi coefficient, computed as follows:

$$M_t$$

		0	1
	0	a_m	b_m
M_{t-1}	1	c_m	d_m

in which the a, b, c, d are frequencies. Then the definition of phi is:

$$\phi_M = \frac{a_m\, d_m - b_m\, c_m}{\sqrt{(a_m + c_m)(b_m + d_m)(a_m + b_m)(c_m + d_m)}}$$

The computation of ϕ_I is analogous to this computation.

They reported that Tavare (1983) showed that for larger two-way tables the X^2 tests of cross-contingency are unbiased if at least one of the interactant's data is seriously independent. We will refer to the following statistic as z_T, where the subscript stands for Tavare:

$$z_T = \sqrt{\frac{1 - \phi_M\, \phi_I}{1 + \phi_M\, \phi_I}}\; X_1^2$$

Another unexplored solution to the problem, using time-series analysis, is given in the next section.

14.5 Untested Suggestions

14.5.1 Binary Time-Series Analysis Solution

Gottman and Ringland (1981) derived a general solution to the problem of inferring cross relationship between two time series, X_t and Y_t.

First autocorrelation is controlled in one series, X_t, by subtracting out how much can be predicted from the past of X_t. Then the past of Y_t is employed to account for additional variance in the residual. Next, the process is repeated for Y_t. Four possibilities exist and the significance of each can be assessed ($X_t \rightarrow Y_t$, $Y_t \rightarrow X_t$, $X_t \nrightarrow Y_t$, and $Y_t \nrightarrow X_t$). The significance of each relationship can be assessed. The relationship $X_t \rightarrow Y_t$ and $Y_t \rightarrow X_t$ is a bidirectional relationship.

The analysis can be performed by program BIVAR in the Gottman-Williams package (Williams & Gottman, 1981). A description of the mathematics of this analysis is presented in Gottman (1981). The mathematics of the Gottman-Ringland procedure have been applied only to continuous, not binary, time series. It is not known how well they would do with binary data. Kedem (1980) has shown that the spectrum can be generalized for binary data, but it is not currently known how well multivariate generalization from continuous to binary data would do.

14.5.2 Controlling for Autocontingency of More Than One Time Lag

This section outlines a generalization of a procedure based on time-series analysis and applied to categorical data. It summarizes a paper by Gottman (1980). These suggestions are tentative.

Suppose we have to code categories X_t and Y_t, which are either 1 or 0 depending on whether X or Y is detected at time t. The covariance between the two time series (continuous or binary) at lag k is defined as

$$C_{XY}(k) = \frac{1}{N-k} \sum_{t=1}^{N-k} (X_t - \bar{X})(Y_{t+k} - \bar{Y}) \ .$$

If we denote by p_X and p_Y the sample estimates of the unconditional probabilities of occurrence of X and Y, respectively, and by $p_k(Y \mid X)$ the conditional probability at lag k, then it can be shown that (Gottman, 1980):

$$C_{XY}(k) = p_X(p_k(Y \mid X) - p_Y)$$

a test of independence is thus equivalent to a test that $p_k(Y \mid X) - p_Y = 0$, which is our familiar definition of sequential connection.

We can estimate the variance of the covariance using a well-known expression (see, for example, Jenkins & Watts, 1968, p. 338):

$$var\,(C_{XY}(k)) = \frac{1}{N-k} C_X(o)C_Y(o) + \frac{2}{N-k} \sum_{s=1}^{\infty} C_X(s)C_Y(s) \ .$$

For the binary case, it is easy to show that

$$C_X(s) = p_X(p_s(X \mid X) - p_X)$$

$$C_Y(s) = p_Y(p_s(Y \mid Y) - p_Y) \ .$$

Thus,

$$C_X(o) = p_X(1 - p_X)$$

$$C_Y(o) = p_Y(1 - p_Y)$$

and we have

$$var\,(C_{XY}(k)) = \frac{1}{N-k}\, p_X p_Y (1 - p_X)(1 - p_Y) + e_k$$

where[1]

$$e_k = \frac{2}{N-k} \sum_{s=1}^{\infty} C_X(s)C_Y(s) \ .$$

The e_k term measures the effects of autocontingency on the variance of the covariance

$$e_k = \frac{2}{N-k}\, C_X(o)C_Y(o) \sum_{1}^{\infty} r_X(s)r_Y(s)$$

where $r_X(s)$ and $r_Y(s)$ are the autocorrelations.

$$e_k = \frac{2}{N-k}\, p_X(1 - p_X)p_Y(1 - p_Y) \times$$

$$\sum_{1}^{\infty} \frac{(p_k(X \mid X) - p_X)}{1 - p_X} \frac{(p_k(Y \mid Y) - p_Y)}{1 - p_Y}$$

---

[1]The sum usually extends from minus infinity to plus infinity, and since $C_X(-S) = C_X(S)$, the sum is twice the sum from 1 to infinity.

The sum in the last expression can be used to correct $var\ C_{XY}(k)$ for the presence of autocontingency of more than one time lag. The z-score between the two codes X and Y is

$$z_{XY}(k) = C_{XY}(k) / \sqrt{var\ C_{XY}(k)}$$

which can be asymptotically compared to a standard normal distribution under the null hypothesis of independence. The remainder of this discussion estimates the size of the correction.

To estimate this correction we can use the well-known result (see Gottman, 1981) that the autocorrelation function of a stationary time series decreases within an exponential envelope. In this case we can use the first-order autoregressive process approximation and recall that $\sum_{1}^{\infty} (ab)^s = ab / (1 - ab)$. Then it follows that

$$var\ C_{XY} \cong \frac{1}{N-k}\ p_X\ p_Y(1-p_X)\ (1-p_Y)\left[1 + \frac{2ab}{1-ab}\right]$$

$$= \frac{1}{N-k}\ p_X\ p_Y(1-p_X)\ (1-p_Y)\left[\frac{1+ab}{1-ab}\right]$$

where

$$a = (p(X \mid X) - p_X) / (1 - p_X)$$

$$b = (p(Y \mid Y) - p_Y) / (1 - p_Y) .$$

If we use the approximations $\dfrac{1}{1-ab} \cong 1 + ab$ for ab small and $(ab)^2 \cong 0$ for ab small, then

$$e_{XY}(k) \cong \frac{1}{N-k}\ p_X\ p_Y(1-p_X)\ (1-p_Y)\ (1+2ab)$$

Kedem (1980) presented an estimate of each of these correction factors (which are first-order autoregressive parameters) in terms of the number of runs of 1's (more than one 1) in each series:

$$cos \, [2\pi \, (\# \text{ of } 1\text{–runs}) \, / \, (N -1)] \; .$$

Kedem also derived an expression for the *p-th* order autoregressive coefficients for a binary time series (see Appendix 14.1). Using these estimates, it may be feasible to extend the Gottman-Ringland (1981) procedure to binary time series, which would afford us the advantage of being able to assess directionality of influence controlling for autocontingency of longer lags.

Appendix 14.1

Sackett's Computational Correction

In general, if there are k codes, $A_1, A_2, \ldots, A_k$ and we wish to assess the significance of $p(A_1 \mid A_2)$, we would compare it to $p(A_1)$, the unconditional probability. Sackett's suggestion is to modify $p(A_1)$ to $\left[p(A_1) - \dfrac{p(A_2 \mid A_2)}{k - 1} \right]$. We could also average the autocontingency over lags for some analyses.

For example, if we have two speakers, N and X, and we evaluate the lag-one conditional probability, $p(X \mid N)$ (i.e., N has just spoken), we would modify $p(X)$ by subtracting $\dfrac{p(N \mid N)}{2 - 1}$. This is similar to the speakership example just reviewed.

15

RECENT ADVANCES: A BRIEF OVERVIEW

Since the late 1970s, a number of papers have appeared that advance the statistics of sequential analysis. This chapter briefly introduces the reader tp some of them.

15.1 Introduction

Gottman (1979a) defined the concept of *dominance* in marital interaction as asymmetry in predictability of behavior from one member of a dyad to the other. In a dyad, for example, if a husband's behavior at some point in time is better predicted from the wife's behavior at previous points in time than the reverse (i.e., predicting the wife's behavior from the husband's past behavior), then the wife is said to be dominant. Gottman and Ringland (1981) expanded on this notion in the context of mother-infant interaction; they also added the modification that controls for autocorrelation. In this modification, the wife's past behavior needs to add information in predicting the husband's behavior over and above the information provided by the husband's past behavior. The use of "dominance" may be further qualified by the content of the behavior itself; for example, in playful mother-infant interaction the fact that the mother's behavior is highly predicted by the infant's behavior but not conversely may reflect the infant's limited developmental social-cognitive abilities and the mother's responsiveness, and not the infant's dominance.

The focus of studies in dyadic social interaction has recently shifted from testing for such lagged dominance toward formally "discovering" some methodology for modeling the nature and structure of this dependence. Most recent works have been concerned with establishing the direction of the dependence and whether the dyad exhibits mutual dependency or is dominated by one member. In this literature the terms "actor," "partner," "participant," and "member of the dyad" are used interchangeably.

15.2 Kraemer and Jacklin (1979); Mendoza and Graziano (1982); and Iacobucci and Wasserman (1987)

The analysis of contingencies arising from dyadic social interaction has to deal with the issue of statistical interdependency of the observations. Kraemer and Jacklin (1979) reviewed the statistical procedures commonly used in the analysis of dyadic data, noted the drawbacks of each approach, and proposed a procedure designed to eliminate the problems of statistical dependency. Their alternative method is essentially an extension of the matched-pair *t*-test approach to determining whether variables such as sex affected the behavior of the dyad. Although this represented an improvement over the more commonly used techniques at the time, it was a univariate procedure and could be applied only to the analysis of a single dependent variable.

Mendoza and Graziano (1982) extended the Kraemer and Jacklin approach to multivariate relations. They took a different approach to handling the correlations among various behavior category summary measures. They used multivariate analysis of variance (MANOVA), testing for overall group differences in total variance of all behavior categories before employing univariate ANOVAs to individual categories. They also discussed the advantages, applications, and guidelines for the use of the multivariate model. Their paper dealt with some important issues, such as multiple significance tests with many highly correlated dependent variables.

However, these methods employ standard multivariate statistical techniques for continuous data and are not generally appropriate for discrete interaction data. Iacobucci and Wasserman (1987) presented an alternative approach to the multivariate analysis of social interactions. They realized that many interactions or relations yield discrete valued data and are thus better modeled by employing methods for the analysis of categorical data. They described recent methodological advances for social network analysis due to Holland and Leinhardt (1981) and Fienberg and Wasserman (1981) (see also Fienberg, Meyer, & Wasserman [1985] and Wasserman [1987] for mathematical details). The goal of their approach to social network data is to understand the group structure better. This may involve questions such as "What effects do important background variables have on actor behavior?" and "Which actor in the group is most dominant?" They present methods for assessing the effects of attribute variables (such as the age of the partner) on the social network relationships. Their article ends by comparing their methods with those of Kraemer and Jacklin and Mendoza and Graziano. First, they noted, they employed models suited for inherently discrete (and ordered) data instead of treating such observational counts as continuous. Second, their models do not make any assumptions about normally distributed variables but, rather, rely on asymptotic chi-squared distributions for assessing the statistical significance of likelihood ratio test statistics. Third, their models can focus on any subgrouping of group members, for example, males and females as separate subgroups, while also adjusting for individual differences. Fourth, in the partitioning of individuals into separate subgroups, in the Kraemer and Jacklin methodology the subgroups are defined by a dichotomous variable, whereas in the Iacobucci and Wasserman methodology the variable may be polytomous (e.g., the race of the actor). One disadvantage of the Iacobucci and Wasserman models is that they can become extremely complicated when one simultaneously studies more than three relational variables. The mathematics involved can become complex with an abundance of available free parameters to be estimated.

Within this genre of multivariate model building, another related area of interest is the *social relations model* proposed by Kenny (1981) and Kenny and LaVoie (1984), and used by many researchers (see Ingraham & Wright, 1986, and Wright & Ingraham, 1985).

15.3   Wampold and Margolin (1982) and Wampold (1984)

Wampold and Margolin (1982) demonstrated that one may study sequential dyadic interactions and the presence of a dominance effect using nonparametric methods, such as runs tests and Hubert's quadratic assignment paradigm (see Hubert & Schultz, 1976, Hubert & Baker, 1977, and especially Baker & Hubert, 1981). Many of these methods are based on straightforward probability assumptions and well-known parametric tests and may be used by researchers without a sophisticated statistical background. These strategies are applied to transition frequency matrices in which the behavioral states of both members of the dyad are conceptualized as a single sequence. Wampold (1984) subsequently extended these methods and developed tests for dominance. The methods may be useful for initial exploratory analysis prior to model building.

15.4 Dillon, Madden, and Kumar (1983)

Dillon et al. (1983) took an alternative approach toward the analysis of sequential categorical data. Their focus was primarily on model building and testing a wide variety of common hypotheses about the structure of influence in dynamic social interactions.

They demonstrated how an alternative class of modeling techniques commonly called *latent structure analysis* (or latent class modeling) can be applied to contingency tables of dyadic interaction to study dominance and lagged dependence across populations of dyads.

Latent structure analysis was developed by Lazarsfeld and Henry (1968) in an entirely different context. It was later extended by Goodman (1974a, 1974b) to the study of qualitative data. The practice of latent class analysis was then greatly enhanced by the work of Clogg (1979, 1981a, 1981b), who implemented the maximum likelihood procedures for the parameters in latent class structures developed by Goodman. Clogg produced a computer program called "MLLSA" (Maximum Likelihood Latent Structure Analysis) that can be used to test a wide variety of latent structure models. This method, which may be deemed a special case of log-linear models, allows one to analyze polytomous data, incorporate explicit tests of significance for rival causal hypotheses, and account for the status of the observed variables or indicated latent unobservable factors. The interested reader may note that this is an analog of structural equations modeling used with continuous variables (see, for example, Golberger & Duncan, 1973, Aigner & Goldberger, 1977, Joreskog & Sorbom, 1979, Joreskog & Wold, 1982).

Dillon et al. used these methods to investigate the lagged dependence between two partners in a dyad and the dependency across different populations, and to study the issues of dominance and autodependence in dyadic interaction sequences. They employed formal test statistics to select from an array of restricted and unrestricted latent class models fit to various dyadic interaction data.

15.5 Feick and Novak (1985)

Feick and Novak (1985) showed how to model contingency tables formed from ordinal variables using a whole array of log-linear models designed for the analysis of ordinal, categorical data. For example, affect can be considered ordered if it goes from negative to neutral to positive instead of these three categories being viewed as separate and unordered. Feick and Novak employed these models to utilize the information about the ordering to gain insight about the structure of the table. Applying such models to sequential, categorical data, they demonstrated how one can gain some parsimonious insight into the dependence between actors in a dyadic interaction process. They used the SPSS-X program called LOGLINEAR to reanalyze the same set of parent-child sorting preference data analyzed by Dillon et al. (1983) and originally cited by Jennings and Niami (1981).

15.6   Faraone and Dorfman (1987)

Faraone and Dorfman (1987) discussed lag sequential analysis methodology to explore and summarize cross-dependencies in complex interaction sequences. They clarified some confusions about the z-statistics proposed by Sackett and recommended the use of the Allison and Liker (1982) and Gottman (1980) z-statistic as the correct statistic under the assumption of no autodependence. Next they presented a statistical test of cross-dependence that derives from the assumption that each behavioral sequence is a first-order Markov chain and presented a so-called Markov z-statistic appropriate for testing cross-dependency under special conditions of autodependence. Finally, they introduced two robust methods for the analysis of lag sequential data. These were the jackknife and data-split techniques for testing cross-dependence and for estimating confidence intervals about indices of cross-dependence. The advantage of such methods derives from the lack of any strong assumptions about the stochastic process that underlies the social interaction sequence. On the other hand, these methods sacrifice some power.

15.7   Budescu (1984)

Allison and Liker (1982) proposed that dyadic social interaction be modeled by representing behavior sequences in a contingency table, where behaviors for both partners in a dyad are coded for times t and $t + 1$ (i.e., antecedent and consequent events). This made it possible to employ log-linear and logit-linear methods to describe sequential dyadic interaction data and to test various hypotheses about patterns of behavior.

Budescu (1984) found the Allison and Liker treatment helpful but opted to take an entirely different approach in his analysis. He criticized Allison and Liker on three separate counts. First, he questioned the validity of their approach at a fundamental methodological level. The basic unit of observation and sampling in this situation, he argued, is the dyad and not one of its members. The actors within a given dyad are not usually independent, and so the members of a dyad cannot be assumed to constitute a random sample from a large population of unrelated individuals. Because one requirement of data analysis is that the unit of analysis be the unit of randomization (and of observation), any parameter estimation and statistical inference regarding a population of dyads must be based on the analysis of a random sample of dyads. Budescu derived methods to operate on the dyad as the unit of observation.

Budescu's second criticism was that Allison and Liker postulated a common model for both members of the dyad and then estimated one set of parameters for each member. Thus, the parameters in the set are common to both members. Budescu noted that some effects, such as autodependence, may not be common for both members. His modeling methodology permits flexibility; for example, one can model a situation in which the behavior of the husband at time $t + 1$ is a function of the behavior of the husband himself and of the wife at time t, but the behavior of the wife is solely a function of her husband's prior behavior.

Third, Budescu claimed that his method should be preferred on statistical grounds, since his methodology included tests that can be used to identify the best fitting model. Note, however, that Allison and Liker, had they chosen to do so, could also have proposed such tests simply by utilizing standard goodness-of-fit statistics for log-linear models for multidimensional contingency tables.

Budescu adopted Allison and Liker's contingency table formulation. He then described the two primary approaches to the analysis of multivariate contingency table data: maximum likelihood estimation of log-linear models (e.g., Bishop et al., 1975) and weighted least squares estimation of the logit-linear model (e.g., Grizzle, Starmer, & Koch [1969], usually called the GSK procedure). Budescu employed the GSK procedure to his four-way dyadic tables because the models and computations closely resemble usual techniques of regression and analysis of variance (ANOVA) used for continuous multivariate data.

15.8 Log-Linear or Logit-Linear Models?

A controversy has sprung up between the use of log-linear versus logit-linear models.

In the log-linear method the traditional distinction between dependent (or criterion) and independent (or predictor) variables is dropped.

Instead, one models the probability of an observation falling into any given cell of a multidimensional table as a function of main effects and interaction effects of the various factors that define the table. The crux of the estimation lies in the application of a numerical procedure called "iterative proportional fitting" to derive maximum likelihood estimates of different models. MLEs can be obtained for a full hierarchy of models at varying levels of complexity, with or without specific main effects or interaction effects. Goodness-of-fit statistics are computed using traditional Pearson chi-squares or likelihood ratio chi-squares.

The logit-linear method, a generalization of a multivariate general linear modeling approach to categorical data, maintains the distinction between independent and dependent variables. Bock (1975) assumed that the responses follow a multivariate logistic distribution and proposed the use of the Newton-Raphson estimation procedure to obtain MLEs of the unknown parameters. The parameters are really coefficients similar in nature to regression beta weights or effects estimates in analysis of variance. The researcher estimates the parameters and then derives expected frequencies in each cell under the model of interest and tests the goodness of fit of the model. An alternative version of this method, introduced by Grizzle et al. (1969) in a classic paper, provides a general and flexible method for fitting linear models to categorical response probabilities. The dependent variable to be explained is the probability of a particular response or outcome. The main effects and interactions are specified in a given model through the manipulation of a design matrix of effect-coded dummy variables. The researcher normally employs the weighted least squares (WLS) technique to obtain a solution called the "minimum logit chi-square solution." The procedure enables the researcher to construct and estimate nonhierarchical as well as hierarchical models.

Budescu (1984) argued that the GSK procedure should be "adopted and routinely used by researchers in the behavioral sciences because of its close resemblance to ANOVA and regression analysis, which are well-known and understood by workers in their fields." This is indeed an advantage of the logit-linear method, as well as its ability to estimate nonhierarchical models. However, two major drawbacks of the GSK method are that the mathematics involved are quite complicated and the handling of empty (or zero) cells becomes somewhat problematic.

Nevertheless, Feick and Novak (1985) employed log-linear models and argued that the widespread availability of the SPSS-X LOGLINEAR program should make such methods convenient for all researchers.

Iacobucci and Wasserman (1987) also preferred the log-linear method for the same reasons and referred to the growing popularity of these methods among social scientists. They pointed out that the only widely available software for implementing the GSK estimation algorithm is the SAS procedure called *Proc Catmod*, whereas a variety of packages can be used to perform log-linear modeling. They also argued that the model structure assumed by GSK may not always be the most appropriate way to view categorical data. We refer the reader to their paper for details.

The controversy will, no doubt, be resolved by the experiences of researchers in applying both methods to a wide assortment of situations.

16

A BRIEF SUMMARY

This chapter summarizes our *major* recommendations as a flow chart for sequential analysis, skipping over many discussions and suboptions already covered in the book.

16.1 The Data

We assume that the data are put in the form of a sequence of numbers representing the codes in our observational category system, times the number of people who are interacting. For example, if we are observing marital interaction and our two codes are "Nice" and "Nasty," we will have four summary numbers, such as, 1 = husband nice, 2 = husband nasty, 3 = wife nice, 4 = wife nasty. The data then might be a stream of numbers such as

1 3 3 3 1 1 1 4 2 4 2 2 2 4 4 1 3 1 1 1 1
1 1 1 1 1 1 3 3 2 4 4 4 4 4 2 2 4 4 1, etc.

16.2 Forming the Timetable: Determining Order and Stationarity

We first wish to make a determination about the order of the Markov chain that would form *the timetable* in each cell of our contextual design, and the stationarity of the data. How this is done is up to the experimenter, but we would suggest doing this globally for all the data combined. Once order and stationarity are determined, the timetable can be formed.

16.3 Homogeneity: The Decisions of How to Best Group Subjects

We have investigated the notion that assessments of the homogeneity of the transition frequencies across subjects (marriages, families, groups) can be an opportunity for exploring the data.

16.4 Within-Subject Analyses

We have also discussed two types of subject-by-subject analyses. The first was to do sequential analyses within each subject and form indices of sequential connection (we reviewed various options). Then these indices can be employed to do standard statistics with the contextual design variables, such as regression or MANOVA. The second option discussed (in Chapter 13) was logistic regression without grouping.

16.5 Pooling Across Subjects

This option is particularly attractive for the study of rare events as well as an overall look at the data. We reviewed log-linear and logit models and logistic regression, talked about how to select a "best" model, how to compare cells of the design, and how to analyze residuals.

16.6 Doing the Minimum

A concern of ours in this book is that the researcher who wishes to use sequential analysis techniques will be overwhelmed by the statistical techniques that are now available. In addition, new methods will likely be developed in the future now that creative social statisticians have seen the importance of this application.

Many researchers will ask, "How can I do these analyses without being overwhelmed by statistical options and mountains of computer output that are difficult to assimilate?"

Here are some suggestions. First, most Markov models in the published research literature show no more than digram structure, so probably researchers' data can be represented with two-way timetables. Table 11.6 affords an example of the kind of table they might wind up with, indicating six codes for marital interaction ($H+$, $H0$, $H-$, $W+$, $W0$, $W-$) and a timetable that is a two-way table of the frequencies of antecedent-by-consequent codes. There is one such timetable for each cell of the contextual design, which is a 2×2 design (one factor is marital happiness and the other is social class).

To analyze this table researchers should learn to use a program like BMDP4F (in the BMDP series) or SPSS's HILOG, and should also learn to do contrasts for log-linear analysis (see Chapters 10 and 11).

These techniques will serve most researchers reasonably well. An additional set of techniques for exploring sequential connections involves lag sequential analysis, which can be done subject by subject or with grouped data. Roger Bakeman's ELAG program will meet most researchers' needs in this regard.

REFERENCES

Agresti, A. (1984). *Analysis of ordinal categorical data*. New York: Wiley.

Aigner, D. J., & Goldberger, A. S. (Eds.). (1977). *Latent variables in socio-economic models*. Amsterdam: North Holland.

Allison, P. D., & Liker, J.K. (1982). Analyzing sequential categorical data on dyadic interaction. *Psychological Bulletin, 91*, 393-403.

Altmann, S.A. (1965). Sociobiology of Rhesus monkeys. II: Stochastics of social communication. *Journal of Theoretical Biology, 8*, 490-522.

Anderson, T. W., & Goodman, L. A. (1957). Statistical inference about Markov chains. *Annals of Mathematical Statisics, 28*, 89-110.

Argyle, M., Clarke, D., & Collert, P. (1980). Project on the sequential structure of social behavior. Final report to SSRC. Unpublished manuscript, University of Oxford.

Arundale, R. B. (1976). *Sampling intervals and the study of change over time in communication: A guideline, data, and implications*. Portland, OR: International Communication Association.

Arundale, R. B. (1982). User's guide to SAMPLE and TEST: Two Fortran IV computer programs for the analysis of discrete state time-varying data, using Markov chain techniques. Fairbanks, Alaska, Department of Speech and Drama: University of Alaska.

Attneave, F. (1959). *Applications of information theory to psychology*. New York: Holt, Rinehart & Winston.

Bakeman, R. (1978). Untangling streams of behavior: Sequential analysis of observational data. In G. P. Sackett (Ed.), *Observing behavior*. Vol. 2: *Data collection and analysis methods*. Baltimore: University Park Press.

Bakeman, R. (1983). Computing lag sequential statistics: The ELAG program. *Behavior Research Methods and Instrumentation, 15*, 530-535.

Bakeman, R., & Adamson, L. B. (1983). Coordinating attention to people and objects in mother-infant and peer-infant interaction. Paper presented at the Society for Research in Child Development meeting in Detroit, April 1983.

Bakeman, R., & Brown, J. V. (1977). Behavioral dialogues: An approach to the assessment of mother-infant interaction. *Child Development, 49*, 195-203.

Bakeman, R., & Brown, J. V. (1980). Analyzing behavioral sequences: Differences between preterm and full-term infant-mother dyads during the first months of life. In D. B. Sawin, R. C. Hawkins, L. O. Walker, & J. H. Penticuff (Eds.), *Exceptional infant* (Vol. 4). New York: Brunner/Mazel.

Bakeman, R., & Dabbs, J. M., Jr. (1976). Social interaction observed: Some approaches to the analysis of behavior streams. *Personality and Social Psychology Bulletin, 2*, 335-345.

Bakeman, R., & Dorval, B. (1988). The independence assumption and its effect on sequential analysis. Unpublished manuscript, Georgia State University, Atlanta.

Bakeman, R., & Gottman, J. (1986). *Observing interaction: An introduction to sequential analysis.* Cambridge University Press.

Baker, F. B., & Hubert, L. J. (1981). A nonparametric technique for the analysis of social interaction data. *Sociological Methods and Research, 9,* 339-361.

Baker, R. J., & Nelder, J. A. (1978). *The GLIM system manual Release 3: Generalized linear interactive modeling.* Oxford, England: The Numerical Algorithms Group (N.A.G.)/Royal Statistical Society.

Bard, Y. (1974). *Nonlinear parameter estimation.* New York: Academic Press.

Barrett, D.E. Significance testing in sequential dependency analysis: Comparison of binomial and multinomial models. Unpublished manuscript.

Bartholomew, D. J. (1982). *Stochastic models for social processes* (3rd ed.). Chichester, England: Wiley.

Bartlett, M. S. (1951). The frequency goodness of fit test for probability chains. *Proceedings of the Cambridge Philosophy Society, 47,* 86-95.

Bass, F. M., Givon, M. M., Kalwani, M. U., Reibstein, D., & Wright, G. P. (1984). An investigation into the order of the brand choice process. *Marketing Science, 3,* 267-287.

Bateson, G., Jackson, D. D., Haley, J., & Weakland, J. (1957). Toward a theory of schizophrenia. *Behavioral Science, 1,* 251-264.

Bekoff, M. (1977). Quantitative studies of three areas of classical ethology: Social dominance, behavioral taxonomy, and behavioral variability. In B. A. Hazlett (Ed.), *Quantitative methods in the study of animal behavior.* New York: Academic Press.

Bendetti, J. K., & Brown, M. B. (1978). Strategies for the selection of log-linear models. *Biometrics, 34,* 680-686.

Berkson, J. (1944). Application of the logistic function to bioassay. *Journal of the American Statistical Assocaition, 39,* 357-365.

Berkson, J. (1951). Why I prefer logits to probits. *Biometrics, 7,* 327-329.

Berkson, J. (1953). A statistically precise and relatively simple method of estimating the bioassay with quantal response based on the logistic function. *Journal of the American Statistical Association, 48,* 565-599.

Berkson, J. (1972). Minimum discrimination information, the "no interaction" problem, and the logistic function. *Biometrics, 28,* 443-468.

Bhapkar, V. P. (1961). Some tests for categorical data. *Annals of Mathematical Statistics, 32,* 72-83.

Bhapkar, V. P. (1966). A note on the equivalence of two test criteria for hypotheses in categorical data. *Journal of the American Statistical Association, 61,* 228-235.

Bhapkar, V. P. (1970). Categorical data analogues of some multivariate tests. In R. C. Bose (Ed.), *Essays in probability and statistics.* Chapel Hill: University of North Carolina Press.

Billings, A. (1979). Conflict resolutions in distressed and nondistressed married couples. *Journal of Consulting and Clinical Psychology, 47,* 365-376.

Billingsley, P. (1961). *Statistical inference for Markov processes.* Chicago: University of Chicago Press.

Billingsley, P. (1961). Statistical methods in Markov chains. *Annals of Mathematical Statistics, 32*, 12-40.

Birch, M. W. (1963). Maximum likelihood in three-way contingency tables. *Journal of the Royal Statistical Society, Series B, 25*, 220-233.

Birch, M. W. (1964). The detection of partial association in the 2×2 case. *Journal of the Royal Statistical Society, Series B, 26*, 313-324.

Birch, M. W. (1965). The detection of partial association II: The general case. *Journal of the Royal Statistical Society, Series B, 27*, 111-124.

Bishop, Y. M. M. (1969). Full contingency tables, logits, and split contingency tables. *Biometrics, 25*, 383-400.

Bishop, Y. M. M., Fienberg, S., & Holland, P. W. (1975). *Discrete multivariate analysis.* Cambridge, MA: MIT Press.

Bliss, C. I. (1935). The calculation of dosage mortality curve (Appendix by R. A. Fisher). *Annals of Applied Biology, 22*, 134-167.

Bobbitt, R. A., Gourevitch, V. P., Miller, E., & Jensen, G. D. (1969). Dynamics of social interactive behavior: A computerized procedure for analyzing trends, patterns, and sequence. *Psychological Bulletin, 71*, 110-121.

Bock, R. D. (1975). *Multivariate statistical methods in behavioral research.* New York: McGraw-Hill.

Bock, R. D., and Yates, G. R. (1973). *Multiqual: log-linear analysis of nominal or ordinal qualitative data by the method of maximum likelihood: User's guide.* Chicago: International Educational Services.

Brazelton, T. B., Koslowski, B., & Main, M. (1974). The origins of reciprocity: The early mother-infant interaction. In M. Lewis & L. A. Rosenblum (Eds.), *The effect of the infant on its caregiver.* New York: Wiley.

Brent, E. E., & Sykes, R. E. (1973). A mathematical model of symbolic interaction between police and suspects. *Behavioral Sciences, 24*, 388-402.

Bresnahan, J. L., & Shapiro, M. M. (1966). A general equation and technique for the exact partitioning of chi-square contingency tables. *Psychological Bulletin, 66*, 252-262.

Brillinger, D. R. (1975). *Time series: Data analysis and theory.* New York: Holt, Rinehart & Winston.

Brillouin, L. (1962). *Science and information theory* (2nd ed.). New York: Academic Press.

Brown, M. B. (1974). Identification of the sources of significance in two-way contingency tables. *Applied Statistics, 23*, 405-413.

Brown, M. B. (1976). Screening effects in multidimensional contingency tables. *Journal of the Royal Statistical Society, Series C, 25*, 37-46.

Budescu, D. V. (1984). Tests of lagged dominance in sequential dyadic interaction. *Psychological Bulletin, 96*, 402-414.

Budescu, D. V. (1985). Analysis of dichotomous variables in the presence of serial dependence. *Psychological Bulletin, 97*, 561-574.

Bush, R. R., & Mosteller, F. (1951). A mathematical model for simple learning. *Psychological Review, 58*, 313-323.

Cairns, R. B. (Ed.). (1979). *The analysis of social interactions: Methods, issues, and illustrations.* Hillsdale, NJ: Erlbaum.

Campbell, A., Converse, P. E., & Rodgers, W. L. (1976). *The quality of American life.* New York: Russell Sage Foundation.

Campbell, D. J., & Shapp, E. (1974). Spectral analysis of cyclic behavior with examples from the field cricket. *Animal Behavior, 22,* 862-875.

Capella, J. N. (1979). Talk-silence sequences in informal conversations. I. *Human Communication Research, 6,* 3-17.

Capella, J. N. (1980). Talk and silence sequences in informal conversations. II. *Human Communication Research, 6,* 130-145.

Capella, J. N. (1981). Mutual influence in expressive behavior: Adult-adult and infant-adult dyadic interaction. *Psychological Bulletin, 89,* 101-132.

Capella, J. N., & Planap, S. (1981). Talk and silence sequences in informal conversations. III: Interspeaker influence. *Human Communication Research, 7,* 117-132.

Carlton, A. G. (1969). On the bias of information estimates. *Psychological Bulletin, 71,* 108-109.

Castellan, N. J., Jr. (1965). On the partitioning of contingency tables. *Psychological Bulletin, 64,* 330-338.

Castellan, N. J., Jr. (1979). The analysis of behavior sequences. In R. B. Cairns (Ed.), *The analysis of social behavior.* Hillsdale, NJ: Erlbaum.

Chatfield, C. (1973). Statistical inference regarding Markov chain models. *Applied Statistics, 22,* 7-20.

Chatfield, C., & Lemon, R. E. (1970). Analyzing sequences of behavioural events. *Journal of Theoretical Biology, 29,* 427-445.

Chernoff, H. (1956). Large sample theory: Parametric case. *Annals of Mathematical Statistics, 27,* 1-22.

Cline, T. R. (1979). A Markov analysis of strangers', roommates', and married couples' conversational focus on their relationships. *The Southern Speech Communications Journal, 45,* 55-68.

Clogg, C. C. (1979). Some latent structure models for the analysis of Likert-type data. *Social Science Research, 8,* 287-301.

Clogg, C. C. (1981a). Latent structure models of mobility. *American Journal of Sociology, 86,* 836-868.

Clogg, C. C. (1981b). New developments in latent structure analysis. In D. J. Jackson & E. F. Borgatta (Eds.), *Factor analysis and measurement in sociological research: A multidimensional perspective.* Beverly Hills, CA: Sage.

Cohen, J. (1977). *Statistical power analysis for the behavioral sciences.* New York: Academic Press.

Colgan, P. W., & Smith, J. T. (1978). Multidimensional contingency table analysis. In P. W. Colgan (Ed.), *Quantitative ethology.* New York: Wiley.

Collet, L. S., & Semmel, M. I. (1976). The analysis of sequential behavior in classrooms and social environments: Problems and proposed solutions. Unpublished manuscript, Indiana University.

Collins, L. (1974). Estimating Markov transition probabilities from micro-unit data. *Applied Statistics, 23,* 355-371.

Condon, W. S., & Ogston, W. D. (1967). A seqmentation of behavior. *Journal of Psychiatric Research, 5,* 221-235.

Condon, W. S., & Samder, L. W. (1974). Neonate movement is synchronized with adult speech: Interactional participation and language acquisition. *Science, 183,* 99-101.

Coulthard, M. (1977). *An introduction to discourse analysis.* London: Longmans.

Cousins, P. C., & Power, T. G. (in press). Quantifying family process: Issues in the analysis of interaction sequences. *Family Process.*

Cousins, P. C., Power, T. G., & Carbonari, J. P. (1986). Power and bias in the z-score: A comparison of sequential analytic indices of contingency. *Behavioral Assessment, 8,* 305-317.

Cousins, P. C., & Vincent, J. P. (1983). Supportive and aversive behavior following spousal complaints. *Journal of Marriage and the Family, 45,* 679-682.

Cox, D. R. (1970). *The analysis of binary data.* London: Methuen.

Cronbach, L. J., Gleser, G. C., Nanda, H., & Rajaratnam, N. (1972). *The dependability of behavioral measurements.* New York: Wiley.

Darroch, J. N. (1962). Interactions in multi-factor contingency tables. *Journal of the Royal Statistical Society, Series B, 24,* 251-263.

Darroch, J. N., & Ratcliff, D. (1972). Generalized iterative scaling for log-linear models. *Annals of Mathematical Statistics, 43,* 1470-1480.

Delius, J. D. (1969). A stochastic analysis of the maintenance behavior of skylarks. *Behavior, 33,* 137-178.

Deming, W. E., & Stephan, F. F. (1940). On a least squares adjustment of a sampled frequency table when the expected marginal totals are known. *Annals of Mathematical Statistics, 11,* 427-444.

Denny, J. L., & Yakowitz, S. J. (1978). Admissible run-contingency type tests for independence and Markov dependence. *Journal of the American Statistical Association, 73,* 177-181.

Dillon, W. R., Madden, T. J., & Kumar, A. (1983). Analyzing sequential categorical data on dyadic interaction: A latent structure approach. *Psychological Bulletin, 94,* 564-583.

Dindia, K. (1981). Reciprocity of self-disclosure: Limitations and illusions. Unpublished manuscript, University of Wisconsin, Milwaukee.

Dixon, W. J. (Ed.). (1985). *BMDP statistical software.* Berkeley: University of California Press.

Domencich, T., & McFadden, D. (1975). *Urban travel demand: A behavioral analysis.* Amsterdam: North Holland.

Douglas, J. M., & Tweed, R. L. (1979). Analyzing the patterns of a sequence of discrete behavioral events. *Animal Behavior, 27,* 1236-1252.

Draper, N. R., & Smith, H. (1981). *Applied regression analysis* (2nd ed.). New York: Wiley.

Dyke, G. V., & Patterson, H. D. (1952). Analysis of factorial arrangements when the data are proportions. *Biometrics, 8,* 1-12.

Ekman, P., & Friesen, W. (1978). *The facial action coding system.* Palo Alto, CA: Consulting Psychologists Press.

Ekman, P., Friesen, W., & Simons, R. (1985). Is the startle reaction an emotion? *Journal of Personality and Social Psychology, 49,* 1416-1426.

Ellis, D. G. (1979). Relational control in two group systems. *Communication Monographs, 46,* 153-166.

Fagen, R. M., & Mankovich, N. J. (1978). Residual analysis and partitioning of social behavior and behavior sequence contingency tables, including significance analysis of two-act transitions in behavioral sequences. In P. W. Colgan (Ed.), *Quantitative ethology*. New York: Wiley.

Fagen, R. M., & Mankovich, N. J. (1980). Two-act transitions, partitioned contingency tables, and the "significant cells" problem. *Animal Behavior, 28*, 1017-1023.

Fagen, R. M., & Young, D. Y. (1978). Temporal patterns of behavior: Durations, intervals, latencies and sequences. In P. W. Colgan (Ed.), *Quantitative ethology*. New York: Wiley.

Faraone, S., & Dorfman, D. (1987). Lag-sequential analysis: Robust statistical methods. *Psychological Bulletin, 101*, 312-323.

Fay, R. E., & Goodman, L. A. (1975). *ECTA program: Description for users*. Chicago: Department of Statistics, University of Chicago.

Feick, L. F., & Novak, J. A. (1985). Analyzing sequential categorical data on dyadic interactions: Log-linear models exploiting the order in variables. *Psychological Bulletin, 98*, 600-611.

Fienberg, S. E. (1969). Preliminary graphical and quasi-independence for two-way contingency tables. *Applied Statistics, 18*, 153-168.

Fienberg, S. E. (1977). *The analysis of cross-classified categorical data*. Cambridge, MA: MIT Press.

Fienberg, S. E. (1980). *The analysis of cross-classified categorical data* (2nd ed.). Cambridge, MA: MIT Press.

Fienberg, S. E., Meyer, M. M., & Wasserman, S. (1985). Statistical analysis of multiple sociometric relations. *Journal of the American Statistical Association, 80*, 51-67.

Fienberg, S. E., & Wasserman, S. (1981). Categorical data analysis of single sociometric relations. In S. Leinhardt (Ed.), *Sociological methodology 1980*. San Francisco: Jossey-Bass.

Fisher, R. A. (1935). Appendix to Bliss. The case of zero survivors. *Annals of Applied Biology, 22*, 164-165.

Fleiss, J. L. (1971). Measuring nominal scale agreement among many raters. *Psychological Bulletin, 76*, 378-382.

Forthofer, R. N., & Lehnen, R. G. (1981). *Public program analysis: A new categorical data approach*. Belmont, CA: Wadsworth.

Gardner, W., & Hartmann, D. P. (1984). On Markov dependence in the analysis of social interaction. *Behavioral Assessment, 6*, 229-236.

Gardner, W., Hartmann, D. P., & Mitchell, C. (1982). The effects of serial dependence on the use of chi-square for analyzing sequential data. *Behavioral Assessment, 4*, 75-82.

Garner, W. R. (1958). Symmetric uncertainty analysis and its implications for psychology. *Psychological Review, 65*, 183-196.

Garner, W. R., & Hake, H. H. (1951). The amount of information in absolute judgements. *Psychological Review, 58*, 446-459.

Gimmelblau, D. M. (1972). *Applied nonlinear programming*. New York: McGraw-Hill.

Ginsberg, D., & Gottman, J. (1986). The conversations of college roommates. In J. M. Gottman & J. Parker (Eds.), *Conversations of friends*. Cambridge University Press.

Gokhale, D. V. (1972). Analysis of log-linear models. *Journal of the Royal Statistical Society, Series B, 34,* 371-376.

Gokhale, D. V., & Kullback, S. (1978). *The information contingency tables.* New York: Marcel Dekker.

Goldberger, A. S., & Duncan, O. D. (Eds.). (1973). *Structural equation models in the social sciences.* New York: Seminar Press.

Goodman, L. A. (1970). The multivariate analysis of qualitative data: Interactions among multiple classifications. *Journal of the American Statistical Association, 65,* 226-256.

Goodman, L. A. (1971a). The analysis of multidimensional contingency tables: Stepwise procedures and direct estimation methods for building models for multiple classifications. *Technometrics, 13,* 33-61.

Goodman, L. A. (1971b). The partitioning of chi-square, the analysis of marginal contingency tables. *Journal of the American Statistical Association, 66,* 339-344.

Goodman, L. A. (1972a). A general model for the analysis of surveys. *American Journal of Sociology, 77,* 1035-1086.

Goodman, L. A. (1972b). A modified multiple regression appraoch to the analysis of dichotomous variables. *American Sociological Review, 37,* 28-46.

Goodman, L. A. (1974a). The analysis of systems of qualitative variables when some of the variables are unobservable. Part I: A modified latent structure approach. *American Journal of Sociology, 79,* 1179-1259.

Goodman, L. A. (1974b). Exploratory latent structure analysis using both identifiable and unidentifiable models. *Biometrika, 61,* 215-231.

Goodman, L. A. (1984). *The analysis of cross-classified data having ordered categories.* Cambridge, MA: Harvard University Press.

Goodman, L. A., & Fay, R. (1973). *ECTA: Everyman's contingency table analysis, program description.* Chicago: University of Chicago Press.

Goodman, L. H., & Kruskal, W. H. (1972). Measures of association for cross classifications, IV: Simplification of asymptotic variance. *Journal of the American Statistical Association, 67,* 415-421.

Gottman, J. M. (1979a). *Marital interaction: Experimental investigations.* New York: Academic Press.

Gottman, J. M. (1979b). Time-series analysis of continuous data in dyads. In M. E. Lamb, S. J. Soumi, & G. R. Stephenson (Eds.), *Social interaction analysis: Methodological issues.* Madison: University of Wisconsin Press.

Gottman, J. M. (1980). On analyzing for sequential connection and assessing interobserver reliability for the sequential analysis of observational data. *Behavioral Assessment, 2,* 361-368.

Gottman, J. M. (1981). *Time-series analysis: A comprehensive introduction for socal scientists.* Cambridge University Press.

Gottman, J. M. (1983). *How children become friends. Monographs of the Society for Research in Child Development* (No. 201).

Gottman, J. M., & Bakeman, R. (1979). The sequential analysis of observational data. In M. E. Lamb, S. J. Soumi, & G. R. Stephenson (Eds.), *Social interaction analysis: Methodological issues.* Madison: University of Wisconsin Press.

Gottman, J. M., Markman, H., & Notarius, C. (1977). The topography of marital conflict: A sequential analysis of verbal and nonverbal behavior. *Journal of Marriage and the Family, 39,* 461-477.

Gottman, M. M., & Mettetal, G. (1986). Speculations sbout social and affective development: Friendship and acquaintanceship through adolescence. In J. Gottman & J. Parker (Eds.), *Conversations of friends: Speculations on affective development.* Cambridge University Press.

Gottman, J. M., & Notarius, C. (1978). Sequential analysis of observational data. In T. Kratochwill (Ed.), *Single subject research.* New York: Academic Press.

Gottman, J. M., & Ringland, J. T. (1981). The analysis of dominance and bidirectionality in social development. *Child Development, 52,* 393-412.

Grizzle, J. E., Starmer, C. F., & Koch, G. G. (1969). Analysis of categorical data by linear models. *Biometrics, 25,* 489-504.

Grizzle, J. E., & Williams, O. D. (1972). Log-linear models and tests of independence for contingency tables. *Biometrics, 28,* 137-156.

Guaagni, P. M., & Little, J. D. C. (1983). A logit model of brand choice calibrated on scanner data. *Marketing Science, 2,* 203-238.

Haberman, S. J. (1973). The analysis of residuals in cross-classified tables. *Biometrics, 29,* 205-220.

Haberman, S. J. (1974). *The analysis of frequency data.* Chicago: University of Chicago Press.

Haberman, S. J. (1978). *Analysis of qualitative data,* Vol. 1: *Introductory topics.* New York: Academic Press.

Haberman, S. J. (1979). *Analysis of qualitative data,* Vol. 2: *New Developments.* New York: Academic Press.

Hanushek, E. A., & Jackson, J. E. (1977). *Statistical methods for social scientists.* New York: Academic Press.

Harper, R. G., Wiens, A. N., & Matarazzo, J. D. (1978). *Nonverbal communication: The state of the art.* New York: Wiley.

Hawes, L. C., & Foley, J. M. (1973). A Markov analysis of interview communication. *Speech Monographs, 40,* 208-219.

Hawes, L. C., & Foley, J. M. (1976). Group decisioning: Testing a finite stochastic model. In G. R. Miller (Ed.), *Explorations in interpersonal communication.* Beverly Hills, CA: Sage.

Hazlett, B. (1980). Patterns of information flow in the hermit crab *Calcinus Tibicen. Animal Behavior, 28,* 1024-1032.

Hewes, D. E. (1975). Statistical tests for Markov chains. Unpublished manuscript, Arizona State University.

Hewes, D. E. (1979). The sequential analysis of social interaction. *The Quarterly Journal of Speech, 65,* 56-73.

Hewes, D. E. (1980). Stochastic modeling of communication processes. In P. Monge & J. Capella (Eds.), *Multivariate techniques in human communication research.* New York: Academic Press.

Hewes, D. E., Planal, P., & Streibel, M. (1980). Analyzing social interaction: Some excruciating models and exhilarating results. In D. I. Nimmo (Ed.), *Communication yearbook IV.* New Brunswick, NJ: Transaction ICA.

Hirokawa, R. Y. (1980). A comparative analysis of communication patterns within effective and ineffective decision-making groups. *Communication Monographs, 47,* 312-321.

Hocking, R. R. (1976). The analysis and selection of variables in linear regression. *Biometrics, 32,* 1-49.

Hoel, P. G. (1954). A test for Markov chains. *Biometrika, 41,*430-433.

Holland, P. W., & Leinhardt, S. (1981). An exponential family of probability densities for directed graphs. *Journal of the American Statistical Association, 76,* 33-51.

Howard, R. (1971). *Dynamic probabilistic systems.* New York: Wiley.

Hubert, L. J., & Baker, F. B. (1977). The comparison and filling of given classification schemes. *Journal of Mathematical Psychology, 16,* 233-253.

Hubert, L. J., & Schultz, J. V. (1976). Quadratic assignment as a general data analysis strategy. *British Journal of Mathematical and Statistical Psychology, 29,* 190-241.

Iacobucci, D., & Wasserman, S. (1987). Dyadic social interactions. *Psychological Bulletin, 102,* 293-306.

Imrey, P. B., Koch, G. G., & Stokes, M. E. (1981). Categorical data analysis: Some reflections on the log linear model and logistic regression, Part I – Historical and methodological overview. *International Statistical Review, 49,* 265-283.

Ingraham, L. J., & Wright, T. L. (1986). A cautionary note on the interpretation of relationship effects in the social relations model. *Social Psychology Quarterly, 49,* 93-97.

Jackson, S. A., & O'Keefe, B. J. (1982). Nonstationary data should not be "corrected." *Human Communication Research, 8,* 146-153.

Jaffe, J., & Feldstein, S. (1970). *Rhythms of dialogue.* New York: Academic Press.

Jenkins, G. M., & Watts, D. G. (1968). *Spectral analysis and its applications.* San Francisco: Holden-Day.

Jennings, K. M., & Niemi, R. G. (1981). *The youth-parent socialization panel study.* Ann Arbor, MI: Intensive Consortium for political and Social Research (ICPSR).

Jöreskog, K. G., & Sörbom, D. (1979). *Advances in factor analysis and structural equations models.* Cambridge, MA: Abt.

Jöreskog, K. G., & Wold, H. (1982). *Systems under direct observation: Causality, structure and prediction, Parts I and II.* Amsterdam: North Holland.

Kedem, B. (1980). *Binary time series.* New York: Marcel Dekker.

Kedem, B., & Slud, E. (1982). Time series discrimination by higher order crossings. *The Annals of Statistics, 10,* 786-794.

Kendall, M., & Stuart, A. (1979). *The advanced theory of statistics,* Vol. 2: *Inference and relationship* (4th ed.). New York: Macmillan.

Kenny, D. A. (1981). Interpersonal perception: A multivariate round robin analysis. In M. B. Brewer & B. E. Collins (Eds.), *Knowing and validating in the social sciences: A tribute to Donald T. Campbell.* San Francisco: Jossey-Bass.

Kenny, D. A., & LaVoie, L. (1984). The social relations model. In L. Berkowitz (Ed.), *Advances in experimental social psychology,* Vol. 18. New York: Academic Press.

Koehler, K., & Larntz, K. (1978). Goodness of fit statistics for large sparse multinomials. Unpublished manuscript, Department of Statistics, University of Minnesota.

Koopmans, L. H. (1974). *The spectral analysis of time series.* New York: Academic Press.

Kraemer, H. C., & Jacklin, C. N. (1979). Statistical analysis of dyadic social behavior. *Psychological Bulletin, 86,* 217-224.

Krokoff, L. J., Gottman, J. M., & Roy, A. K. (1988). Blue collar and white collar marital interaction and communication orientation. *Journal of Social and Personal Relationships, 5,* 201-221.

Ku, H. H., & Kullback, S. (1968). Interaction in multidimensional contingency tables: An information theoretic approach. *Journal of Research – National Bureau of Standards (Mathematical Sciences), 72B,* 159-199.

Ku, H. H., & Kullback, S. (1969). Analysis of multidimensional contingency tables: An information theoretic approach. *Bulletin of the International Statistical Society, 43,* 156-158.

Ku, H. H., & Kullback, S. (1974). Loglinear models in contingency table analysis. *American Statistician, 28,* 115-122.

Ku, H. H., Varner, R. N., & Kullback, S. (1971). Analysis of multidimensional contingency tables. *Journal of the American Statistical Association, 66,* 55-64.

Kullback, S., Kupperman, M., & Ku, H.H. (1962). Tests for contingency tables and Markov chains. *Technometrics, 4,* 573-608.

Lamb, M. E., Soumi, S. J., & Stephenson, G. R. (Eds.). (1979). *Social interaction analysis: Methodological issues.* Madison: University of Wisconsin Press.

Landis, J. R., Stanish, W. M., Freeman, J. L., & Koch, G. G. (1976). A computer program for the generalized chi-square analysis of categorical data using weighted least squares (GENCAT). *Computer Programs Biomedical, 6,* 196-231.

Lazarsfeld, P. F., & Henry, N. W. (1968). *Latent structure analysis.* Boston, MA: Houghton Mifflin.

Lederer, W. J., & Jackson, D. D. (1968). *The mirages of marriage.* New York: Norton.

Lemon, R. E., & Chatfield, C. (1971). Organization of song in cardinals. *Animal Behavior, 19,* 1-17.

Lewis, M., & Rosenblum, L. A. (1974). *The effect of the infant on its caregiver.* New York: Wiley.

Lichtenberg, J. W., & Hummel, T. J. (1976). Counseling as a stochastic process: Fitting a Markov chain model to initial counseling interviews. *Journal of Counseling Psychology, 23,* 310-315.

Losey, G. S., Jr. (1978). Information theory and communication. In P. W. Colgan (Ed.), *Quantitative ethology.* New York: Wiley.

Lubin, D., & Whiting, B.B. (1977). Learning techniques of persuasion: An analysis of interaction sequences. Society for Research in Child Development, New Orleans.

Madansky, A. (1959). Least squares estimation in finite Markov processes. *Psychometrika, 24,* 137-144.

Madden, T. J., & Dillon, W. R. (1982). Causal analysis and latent class models: An application to a communication hierarchy of effects model. *Journal of Marketing Research, 19,* 472-490.

Magidson, J. (1982). Some common pitfalls in causal analysis of categorical data. *Journal of Marketing Research, 19,* 461-471.

Magidson, J., Swan, J. H., & Berk, R. A. (1981). Estimating nonhierarchical and nested log-linear models. *Sociological Methods and Research, 10,* 3-49.

Margolin, G., & Wampold, B. E. (1981). Sequential analysis of conflict and accord in distressed and nondistressed marital partners. *Journal of Consulting and Clinical Psychology, 47*, 554-567.

Marko, H. (1973). The bidirectional communications theory – A generalization of information theory. *IEEE Transactions on Communication, 21*, 1345-1351.

McCullagh, P. (1980). Regression models for ordinal data (with discussion). *Journal of the Royal Statistical Society, Series B, 42*, 109-142.

McCullagh, P., & Nelder, J. A. (1983). *Generalized linear models*. London: Chapman & Hall.

McFadden, D. (1973). Conditional logit analysis of qualitative choice behavior. In P. Sarembka (Ed.), *Frontiers in econometrics*. New York: Academic Press.

Mendoza, J. L., & Graziano, W. G. (1982). The statistical analysis of dyadic social behavior: A multivariate approach. *Psychological Bulletin, 92*, 532-540.

Miller, G. A. (1952). Finite Markov processes in psychology. *Psychometrika, 17*, 149-167.

Miller, G. A., & Frick, F. C. (1949). Statistical behavioristics and sequences of responses. *Psychological Review, 56*, 311-324.

Mjrberg, A. A. (1972). Ethology of the bicolor damselfish, *Eupomaclatsus partitus (Pisces Pomacentridae)*: A comparative analysis of laboratory and field behaviour. *Animal Behaviour Monographs, 5*.

Murstein, B. I., Cerreto, M., & MacDonald, M. G. (1977). A theory and investigation of the effect of exchange-orientation on marriage and friendship. *Journal of Marriage and the Family, 39*, 543-548.

Natale, M. (1976). A Markovian model of adult gaze behavior. *Journal of Psycholinguistic Research, 5*, 53-67.

Nelder, J. A. (1974). Log-linear models for contingency tables: A generalization of canonical least squares. *Applied Statistics, 23*, 323-329.

Nelder, J. A., & Wedderburn, R. W. (1972). Generalized linear models. *Journal of the Royal Statistical Society, Series A, 135*, 370-384.

Nerlove, M., & Press, S. J. (1973). Univariate and multivariate log-linear and logistic models. (Technical Report R-1306-EDA/NIH). Santa Monica, CA: Rand Corporation.

Neter, J., Wasserman, W., & Kutner, K. H. (1985). *Applied linear statistical models: Regression, analysis of variance and experimental designs* (2nd ed.). Homewood, IL: Richard D. Irwin.

Neyman, J. (1949). Contribution to the theory of the chi-square test. In J. Neyman (Ed.), *Proceedings of the First Berkeley Symposium on Mathematical Statistics and Probability*. Berkeley: University of California Press, 230-273.

Norusis, M. J. (1985). *SPSSX: Advanced statistics guide*. New York: McGraw-Hill.

Odoroff, C. L. (1970). A comparison of minimum logit chi-square estimation and maximum likelihood estimation in a 2×2 and $3 \times 2 \times 2$ contingency tables: Tests for interaction. *Journal of the American Statistical Association, 65*, 1617-1631.

Overall, J. E. (1980). Continuity correction for Fisher's exact probability test. *Journal of Educational Statistics, 5*, 177-190.

Palmgren, J. (1987). The Fisher information matrix for log linear models arguing conditionally on observed explanatory variables. *Biometrika, 68*, 563-566.

Patterson, G. R., Littman, R. A., & Bricker, W. (1967). Assertive behavior in children: A step toward a theory of aggression. *Monographs of the Society for Research in Child Development, 32,* 1-43.

Pierce, D. A. (1977). Relationships – and the lack thereof – between economic time-series, with special reference to money and interest rates. *Journal of the American Statistical Association, 73,* 11-26.

Plackett, R. L. (1962). A note on interactions in contingency tables. *Journal of the Royal Statistical Society, Series B, 24,* 162-166.

Plackett, R. L. (1974). *The analysis of categorical data.* London: Griffin.

Plackett, R. L. (1981). *The analysis of categorical data* (2nd ed.). London: Griffin.

Powell, M. J. D. (Ed.). (1982). *Nonlinear optimisation, 1987.* London: Academic Press.

Quastler, H. (1958). A primer on information theory. In H. P. Yockey, R. L. Platzman, & H. Quastler (Eds.), *Symposium on information theory in biology.* New York: Pergamon Press.

Rao, C. R. (1961). Asymptotic efficiency and limiting information. In J. Neyman (Ed.), *Proceedings of the Fourth Berkeley Symposium on Mathematical Statistics and Probability.* Berkeley: University of California Press.

Rao, C. R. (1973). *Linear statistical inference and its applications* (2nd ed.). New York: Wiley.

Raush, H. L. (1965). Interaction sequences. *Journal of Personality and Social Psychology, 2,* 487-499.

Raush, H. L., Barry, W. A., Hertel, R. K., & Swain, M. (1974). *Communication, conflict and marriage.* San Francisco: Jossey-Bass.

Revenstorf, D., Vogel, B., Wegener, C., Halweg, K., & Schindler, L. (1980). Escalation phenomena in interaction sequences. An empirical comparison of distressed and nondistressed couples. *Behaviour Analysis and Modification, 2,* 97-116.

Rhoades, H. M., & Overall, J. E. (1982). A sample size correction for Pearson chi-square in 2 × 2 contingency tables. *Psychological Bulletin, 91,* 418-423.

Rodger, R. S., & Rosebrugh, R. D. (1979). Computing a grammar. *Animal Behavior, 27,* 737-749.

Ross, D. C. (1977). Testing patterned hypotheses in multi-way contingency tables using weighted kappa and weighted chi-square. *Educational and Psychological Measurement, 37,* 291-307.

Sackett, G. P. (1974). A nonparametric lag sequential analysis for studying dependency among responses in observational scoring systems. Unpublished manuscript, University of Washington.

Sackett, G. P. (1978). Measurement in observational research. In G. P. Sackett (Ed.), *Observing behavior,* Vol. 2: *Data collection and analysis methods* (pp. 25-43). Baltimore, MD: University Park Press.

Sackett, G. P. (1979). The lag sequential analysis of contingency and cyclicity in behavioral interaction research. In J. D. Osofsky (Ed.), *Handbook of infant development.* New York: Wiley.

Sackett, G. P. (1980). Lag sequential analysis as a data reduction technique in social interaction research. In D. B. Swain, R. C. Hawkins, L. O. Walker, & J. H. Penticuff (Eds.), *Exceptional infant* (Vol. 4). New York: Brunner/Mazel.

Sandland, R. L. (1976). Application of methods of testing for independence between two Markov chains. *Biometrics, 32,* 629-636.

SAS Institute, Inc. (1985). *SAS User's Guide: Statistics, Version S Edition*. Cary, NC: SAS Institute.

Schaap, C. (1982). *Communication and adjustment in marriage*. Lisse: Svets & Zeitlinger.

Scherer, K. R., & Ekman, P. (Eds.). (1982). *Handbook of methods in nonverbal behavior research*. Cambridge University Press.

Schouwenburg, H. C., Brouwer, H. J., Jorg, T., & Boekhout, C. I. M. (1978). Observing doctor-patient interaction. I. Construction of a category system. Unpublished manuscript.

Schouwenburg, H. C., Brouwer, H. J., Jorg, T., & Boekhout, C. I. M. (1978). Observing doctor-patient interaction. II. Construction of a model. Unpublished manuscript.

Shannon, C. E., & Weaver, W. (1949). *The mathematical theory of communication*. Urbana: University of Illinois Press.

Sillars, A. L. (1980). The sequential and distributional structure of conflict interactions as a function of attributions concerning the locus of responsibility and stability of conflicts. In D. I. Nimmo (Ed.), *Communication Yearbook 4*. New Brunswick, NJ: Transaction ICA.

Simpson, E. H. (1951). The interpretation of interaction in contingency tables. *Journal of the Royal Statistical Society, Series B, 13*, 238-241.

Singer, B., & Spilerman, S. (1975). The representation of social processes by Markov models. *American Journal of Sociology, 82*, 1-54.

Sloane, D., Notarius, C., & Pellegrini, D. S. (1985). Family adaptation to parental dysfunction. Unpublished manuscript, Catholic University, Washington, D. C.

SPSS, Inc. (1986). *SPSSX User's Guide, 2nd edition*. Chicago: SPSS, Inc.

Stokes, M. E., & Koch, G. G. (!980). CATMAX: A SAS macro for fitting log-linear models to contingency tables by maximum likelihood. Unpublished technical report. Chapel Hill: University of North Carolina, Department of Biostatistics.

Strodtbeck, F. L. (1951). Husband-wife interaction over revealed differences. *American Sociological Review, 16*, 468-473.

Stuart, R. B. (1969). Operant-interpersonal treatment for marital discord. *Journal of Consulting and Clinical Psychology, 33*, 675-682.

Taerum, T., Ferris, C., Lytton, H., & Zwirner, W. (1976). Programs for the analysis of dependencies in parent-child interaction sequences. *Behavior Research Methods and Instrumentation, 8*, 517-519.

Tavare, S. (1983). Serial dependence in contingency tables. *Journal of the Royal Statistical Society, Series B, 45*, 100-106.

Tavare, S., & Altham, P. M. (1983). Serial dependence of observations leading to contingency tables and corrections to chi-squared statistics. *Biometrika, 70*, 139-144.

Theil, H. (1970). On estimation of relationships involving qualitative variables. *American Journal of Sociology, 76*, 103-154.

Ting-Toomey, S. (1983). An analysis of verbal communication patterns in high and low marital adjustment groups. *Human Communication Research, 9*, 306-319.

Tronick, E. D., Als, H., & Brazelton, T. B. (1977). Mutuality in mother-infant interaction. *Journal of Communication, 27*, 74-79.

Upton, G. J. G. (1978). *The analysis of cross-tabulated data*. New York: Wiley.

Valentine, K. B., & Fisher, B. A. (1974). An interaction analysis of verbal innovative deviance in small groups. *Speech Monographs, 41*, 413-420.

Vuchinich, S. (1985). Sequencing and social structure in family conflict. *Social Psychology Quarterly, 47,* 217-234.

Wald, A. (1943). Tests of statistical hypotheses concerning general parameters when the number of observations is large. *Transactions of the American Mathematical Society, 54,* 426-482.

Walker, S. H., & Duncan, D. B. (1967). Estimation of the probability of an event as a function of several independent variables. *Biometrika, 54,* 167-169.

Wampold, B. E. (1984). Tests of dominance in sequential categorical data. *Psychological Bulletin, 96,* 424-429.

Wampold, B. E., & Margolin, G. (1982). Nonparametric strategies to test the independence of behavioral states in sequential data. *Psychological Bulletin, 92,* 755-765.

Wasserman, S. (1987). Conformity of two sociometric relations. *Psychometrika, 52,* 3-18.

Wasserman, S., & Iacobucci, D. (in press). Statistical analysis of discrete relational data. *British Journal of Mathematical and Statistical Psychology, 39.*

Williams, E. A., & Gottman, J. M. (1981). *A user's guide to the Gottman-Williams time-series analysis computer programs for social scientists.* Cambridge University Press.

Williams, O. D., & Grizzle, J. E. (1972). Analysis of contingency tables having ordered response categories. *Journal of the American Statistical Association, 67,* 55-63.

Wright, J. C., & Ingraham, L. J. (1985). Simultaneous study of individual differences and relationship effects in social behaviors in groups. *Journal of Personality and Social Psychology, 48,* 1041-1047.

Yakowitz, S. J. (1976). Small sample hypothesis tests of Markov order with application to simulated and hydrologic chains. *Journal of the American Statistical Association, 71,* 132-136.

NAME INDEX

Page numbers in italics indicate material in tables.

SUBJECT INDEX